Research and Perspectives in Alzheimer's Disease

For further volumes:
http://www.springer.com/series/1175

Richard I. Morimoto • Yves Christen
Editors

Protein Quality Control in Neurodegenerative Diseases

Springer

Editors
Richard I. Morimoto
Department of Molecular Biosciences
Rice Institute for Biomedical Research
Northwestern University
2205 Tech Drive, Hogan 2-100
Evanston, IL 60208
USA

Yves Christen
Fondation IPSEN pour la
Recherche Therapeutique
65, quai George Gorse,
92650 Boulogne Billancourt Cedex,
France

ISSN 0945-6066
ISBN 978-3-642-27927-0 ISBN 978-3-642-27928-7 (eBook)
DOI 10.1007/978-3-642-27928-7
Springer Heidelberg New York Dordrecht London

Library of Congress Control Number: 2012953474

Printed on acid-free paper

Springer is part of Springer Science+Business Media (www.springer.com)

Foreword

Protein Quality Control in Neurodegenerative Diseases

Fundamental to all cellular processes is the efficient functioning of the proteome. This is of central importance to the health of the cell and lifespan of the organism. Consequently, each protein is folded and maintained in a functional state by the proteostasis network regulated by cell stress responses and comprised of molecular chaperones, transport machineries, the ubiquitin-dependent proteasome and autophagic activities that function in concert to orchestrate proteome health. Stress and aging, however, challenge the proteostasis network to maintain balance, and in certain cases become limiting thus increasing the risk of cellular pathology and disease.

There is now substantial evidence that the accumulation of damaged proteins not only has direct consequences on the efficiency and fidelity of cellular processes, but when uncorrected initiates a cascade of dysfunction. In humans, this is associated with a plethora of diseases of protein conformation. These include many of the most challenging human ailments including Alzheimer's disease (AD), Parkinson's disease (PD), Amyotrophic lateral sclerosis (ALS), Huntington's disease (HD), metabolic diseases, and cancer. While each disease has a distinctive clinical profile with characteristic tissue vulnerability and age-dependent onset, they all have in common that the expression aggregation-prone proteins that interfere profoundly with cellular function leading to pathology and disease.

The present volume gathers contributions to the Colloque Médecine et Recherche organized by the Fondation IPSEN, in May 9, 2011, in Paris.

Richard I. Morimoto
Department of Molecular Biosciences,
Rice Institute for Biomedical Research, Northwestern University,
Evanston, USA

Yves Christen
Fondation IPSEN, Paris, France

Acknowledgments

The editors wish to express their gratitude to Mrs Mary Lynn Gage and Céline Maffre for her editorial assistance, Mrs Astrid de Gérard, Jacqueline Mervaillie and Sonia Le Cornec for the organization of the meeting.

Contents

Contributors

Christian Behl Institute for Pathobiochemistry, University Medical Center, Johannes Gutenberg University, Mainz, Germany

Ana Maria Cuervo Department of Developmental and Molecular Biology, Department of Medicine and Institute for Aging Studies, Albert Einstein College of Medicine, Bronx, NY, USA

Andy Dillin Molecular and Cell Biology Laboratory, Salk Institute for Biological Studies, La Jolla, CA, USA

Steven Finkbeiner Gladstone Institutes of Neurological Disease, Taube-Koret Center for Huntington's Disease Research, Hellman Family Foundation Alzheimer's Disease Research Program, and the Departments of Neurology and Physiology, University of California, San Francisco, CA, USA

Judith Frydman Department of Biology and BioX Program, Stanford University, Stanford, CA 94305-5020, USA

Jiefei Geng Life Sciences Institute, Departments of Molecular, Cellular and Develop. Biology, University of Michigan, Ann Arbor, MI, USA

Eric Guisbert Department of Molecular Biosciences, Rice Institute for Biomedical Research, 2205 Tech Drive, Hogan 2–100, Northwestern University, Evanston, IL, USA

Susmita Kaushik Department of Developmental and Molecular Biology, Department of Medicine and Institute for Aging Studies, Albert Einstein College of Medicine, Bronx, NY, USA

Daniel J. Klionsky Life Sciences Institute, Deptartments of Molecular, Cellular and Develop. Biology, University of Michigan, Ann Arbor, MI, USA

Melinda A. Lynch-Day Life Sciences Institute, Deptartments of Molecular, Cellular and Develop. Biology, University of Michigan, Ann Arbor, MI, USA

Noboru Mizushima Department of Physiology and Cell Biology, Tokyo Medical and Dental University, Bunkyo-ku, Tokyo Japan

Richard I. Morimoto Department of Molecular Biosciences, Rice Institute for Biomedical Research, Northwestern University, 2205 Tech Drive, Hogan 2-100, Evanston, IL 60208, USA

Christian Neri Neuronal Cell Biology and Pathology, INSERM Unit 894, Paris, France

Ian Nicastro Molecular and Cell Biology Laboratory, Salk Institute for Biological Studies, La Jolla, CA, USA

David Ron Institute of Metabolic Sciences, MRL, level 4, University of Cambridge, Cambridge UK

Koning Shen Department of Biology and BioX Program, Stanford University, Stanford, CA 94305-5020, USA

Cendrine Tourette Buck Institute, Novato, CA, USA

Rafael Vazquez-Manrique Neuronal Cell Biology and Pathology, INSERM Unit 894, Paris, France

Wei-Lien Yen Life Sciences Institute, Departments of Molecular, Cellular and Develop. Biology, University of Michigan, Ann Arbor, MI, USA

The regulation and function of the heat shock response

Eric Guisbert and Richard I. Morimoto

Abstract The heat shock response (HSR) is an essential cellular and organismal protective mechanism against acute forms of physiological and environmental stress. Induction of the HSR by exposure to acute stress conditions such as elevated temperature, oxidants, and heavy metals involves activation of heat shock transcription factors (HSFs) and elevated expression of genes encoding heat shock proteins and molecular chaperones that restore proteostasis and prevent the further accumulation of misfolded and aggregated proteins. Here, we examine the stress-sensing mechanisms that regulate HSFs and the role of post-translational modifications that regulate HSF activity. In addition to its role in acute stress, the expression of molecular chaperones and heat shock genes is important for development and is protective when challenged by chronic proteostasis imbalance during aging and diseases of protein conformation.

Introduction

A universal requirement of all living systems is the ability to sense and respond to diverse forms of environmental and physiological stress and to respond precisely to the type and level of damage. Cellular adaptations to potentially harmful changes in the environment are often referred to as stress responses, and they are defined by the specific stresses to which they respond (e.g., temperature stress, oxidative stress, osmotic stress). The heat shock response (HSR) represents one of the most ancient and highly conserved molecular stress responses.

The HSR detects disruptions of protein homeostasis, or proteostasis, induced by temperature fluctuations, oxidants, and other perturbants of protein expression and

E. Guisbert • R.I. Morimoto (✉)
Department of Molecular Biosciences, Rice Institute for Biomedical Research, Northwestern University, 2205 Tech Drive, Hogan 2-100, Evanston, IL 60208, USA
e-mail: r-morimoto@northwestern.edu

R.I. Morimoto and Y. Christen (eds.), *Protein Quality Control in Neurodegenerative Diseases*, Research and Perspectives in Alzheimer's Disease,
DOI 10.1007/978-3-642-27928-7_1,

folding (Morimoto 2008, 2011; Akerfelt et al. 2010). Many of these stress conditions interfere with protein expression, folding to active states, assembly into molecular machines, and clearance of damaged proteins. The efficient functioning of the proteome requires a highly efficient quality control system, the proteostasis network (PN), which is fundamental to all cellular processes and central to organismal health and lifespan. The HSR, by regulating the expression of molecular chaperones and other components of the PN, senses proteotoxic stress and prevents further protein damage. Activation of the HSR has broad protective properties for the cell and often cross-protects against a severe exposure to other physiological, environmental or chemical stress conditions, and mutations that disrupt the HSR have devastating effects (Zhou et al. 1988; Jedlicka et al. 1997; McMillan et al. 1998; Xiao et al. 1999; Zhang et al. 2002; Hajdu-Cronin et al. 2004).

Induction of the HSR is often just one of multiple molecular events that occur in parallel to minimize protein damage. Activation of the HSR and expression of chaperones have an immediate beneficial effect to prevent misfolding and aggregation; likewise, the HSR-induced attenuation of de novo protein synthesis prevents the expression of nascent polypeptides. HSR-induced translational arrest involves phosphorylation of the translation initiation factor eIF-2a, which prevents recycling of the eIF-2 complex, which in turn inhibits recruitment of the initiating MET-tRNA to the ribosome (Bonanou-Tzedaki et al. 1978; Matts et al. 1983; Hurst and Matts 1987; Matts and Hurst 1989; Mendez et al. 1992; Mendez and de Haro 1994). Inhibition of translation by heat shock also leads to relocalization of mRNAs and many translation and RNA processing factors to specific cytoplasmic foci known as stress granules (Kedersha et al. 1999). Further, HSR attenuates RNA processing via dephosphorylation of SRp68 and perturbation of splicing (Shin et al. 2004; Shi and Manley 2007). Thus, multiple mechanisms function together to reduce the load of protein misfolding in the face of increased temperature, although it remains to be determined whether these processes are independent or are coordinated at some level.

HSR gene induction

The transcription of heat shock (HS) genes is regulated by a family of conserved heat shock transcription factors (HSFs). HSF1 is conserved in humans, Drosophila, *C. elegans*, and yeast whereas HSF2, 3, and 4 are expressed in vertebrates and exhibit diverse functional properties during development, stress, and life span (Pirkkala et al. 2001). Much of our knowledge regarding HSR-dependent transcriptional regulation has come from work focused on activation of the HSP70 gene. Analysis of the HSP70 promoter in human cells identified multiple *cis* elements in addition to heat shock elements (HSEs), which are the binding sites for HSF1 (Greene et al. 1987; Morgan et al. 1987; Wu et al. 1987; Williams et al. 1989; Abravaya et al. 1991). These additional elements include binding sites for NF-Y,

HSP-CBF, and SP1 and are thought to represent basal factors that are required for inducibility (Taylor et al. 1991; Lu et al. 1993). In *Drosophila*, even in the absence of stress, the HSP70 gene is maintained in an open chromatin state and RNA polymerase is recruited to the promoter, initiates transcription and pauses to generate a short nascent transcript (Rougvie and Lis 1988). Upon heat shock, HSF1 binds to the HSE and recruits pTEFb, which releases the paused polymerase, thus allowing rapid induction of HSP70 transcripts (Lis et al. 2000; Boehm et al. 2003). Furthermore, heat shock causes rapid, transcription-independent loss of nucleosomes over the HSP70 gene locus, and HSF, GAGA factor, and PARP are required for this activity (Petesch and Lis, 2008). Interestingly, in yeast, activation of the HSR does not correlate with the recruitment of HSF1 to DNA; instead, HSF1 is constitutively bound (Sorger et al. 1987; Sorger and Pelham 1988).

Characterization of the HSF1 regulon in various model systems has revealed that genes with diverse functions are regulated by heat shock (GuhaThakurta et al. 2002; Hahn et al. 2004; Trinklein et al. 2004; Birch-Machin et al. 2005; Yamamoto et al. 2005; Page et al. 2006). In addition to the role of HSPs as molecular chaperones, HS-induced genes are involved in energy generation, trafficking, cell integrity, transcription, splicing, chromosome organization, DNA repair, anti-apoptosis, ubiquitination, carbohydrate metabolism, transport, cytoskeleton, and the cell cycle. While HSF1 itself is not conserved beyond eukaryotes, the HSR regulon in prokaryotes is similar to that of eukaryotes (Nonaka et al. 2006). A detailed comparison of genes induced by HS in an HSF1-dependent manner, genes that are bound by HSF1 as measured by ChIP, and genes containing HSEs revealed surprisingly mediocre correlations between all three approaches (Trinklein et al. 2004). For example, many HSF-bound promoters did not contain a strong HSE, and many genes with a strong HSE and detectable HSF binding were not heat inducible. One potential explanation is that distinct types of HSF1-binding sites could exist, and indeed there is some evidence for this in yeast (Yamamoto et al. 2005). Additionally, there is evidence that the chromatin landscape has a significant effect on HSF1 binding (Guertin and Lis 2010).

The HSF1 activation pathway

HSF1 is regulated by a multistep process initiated by the conversion of the inert monomeric HSF1 to nuclear-localized HSF1 homotrimers, followed by binding to HSEs upstream of HS genes and leading to elevated transcription of HS genes (Fig. 1). In the absence of stress, human HSF1 exists in either the cytoplasm or nucleus as an inactive monomer. In this state, HSF1 is post-transcriptionally modified by phosphorylation on at least three serine residues (S303, S307, and S363) that helps maintain it in the repressed state (Sorger and Pelham 1988; Chu et al. 1996; Knauf et al. 1996; Kim et al. 1997; Kline and Morimoto 1997; He et al. 1998; Dai et al. 2000; Holmberg et al. 2001; Guettouche et al. 2005). Upon HS, HSF1 rapidly transitions to a transcriptionally active trimer that localizes within the

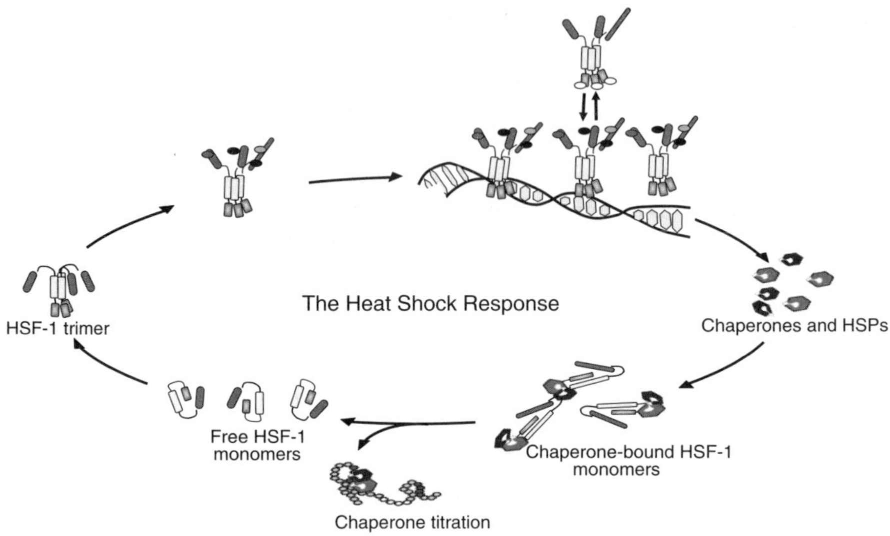

Fig. 1 Transcriptional activation of the heat shock response (HSR). Multistep pathway for the regulation of HSF1 is initiated from the inert, chaperone-associated monomeric state present in control, non-stressed cells. Upon exposure of cells to environmental and physiological stress conditions (heat shock, oxidative stress), the chaperones are titrated by the flux of non-native proteins, leading to HSF1 trimerization, post-translational modification by phosphorylation, sumoylation, and acetylation that affect transcriptional activity and DNA binding. The active HSF1 trimers bind to the heat shock elements in the promoter regions of target HS genes, resulting in the elevated the expression of chaperones and other heat shock proteins (HSPs)

nucleus to regulate the expression of HS genes. Activation of HSF1 by heat shock also leads to hyperphosphorylation of HSF1 at serine residues such as S230 that affect transcriptional activity (Sorger et al. 1987; Larson et al. 1988; Sorger and Pelham 1988; Chu et al. 1996, Cotto et al. 1996; Chu et al. 1998; Xia et al. 1998; Bijur and Jope 2000; Holmberg et al. 2001). The kinases that have been implicated in HSF regulation include CaMKII, GSK-3a, PKCa, ERK, and JNK; however, it remains to be demonstrated how these kinases become stress responsive and regulate HSF1 activity.

Another post-translational modification of HSF1 is SUMOylation on lysine 298, which is triggered by a preceding phosphorylation event at S303 (Hietakangas et al. 2003; Anckar et al. 2006). The kinetics of SUMOylation mimic the kinetics of HSF activation; however, the extent of SUMOylation negatively correlates with the extent of stress. These data suggest that SUMOylation may negatively regulate HSF1 during mild stress.

Upon activation, HSF1 is recruited to nuclear-localized stress granules in human cells (Sarge et al. 1993). These granules are distinct from other types of foci such as P bodies, Cajal bodies, PML bodies, etc. The formation of stress granules correlates with recruitment of HSF1 to satellite repeat regions of DNA, where it activates the transcription of non-coding RNAs (Denegri et al. 2002; Jolly et al. 2002, 2004;

Allen et al. 2004; Rizzi et al. 2004; Mariner et al. 2008). These non-coding RNAs have been proposed to have a role in general RNA PolII-mediated transcription.

The attenuation of the HSR involves the transcriptional silencing of HSF1 bound to the HSE, the release of HSF1 trimers from the HSEs, dephosphorylation of HSF1, and conversion of trimers back to the inert monomer (Wu 1995; Morimoto 1998; Westerheide and Morimoto 2005; Anckar and Sistonen 2007). This step of the HSR is associated with the reversible acetylation of HSF1 on 9 lysine residues, one of which, K80, occurs in the DNA binding domain and prevents the interaction with DNA (Westerheide et al. 2009). HSF1 acetylation at K80 is regulated by multiple histone acetyltransferases, including p300/CBP and the NAD-dependent deacetylase, SIRT1. Consequently, the regulation of the HSR by SIRT1 provides a functional link between the HSR, metabolic stress, and longevity.

While extensive work has revealed many features of the HSR, many of the details of the regulatory mechanisms are lacking. The HSF1 cycle is comprised of multiple steps, many of which have been identified, yet how these steps are coordinated is largely unexplored. Such a plethora of regulatory modifications would allow for a wide range of HSF1 regulatory states to accommodate the requirements for HSF1 under both non-stress and stress conditions. However, a description of the true nature of existing regulatory states will require the development of new tools and approaches.

Regulation of the HSR

Maintenance of HSF1 in its control, inert state involves weak and transient interactions with the chaperones Hsp90, Hsp70 and Hsp40 (Abravaya et al. 1992; Ali et al. 1998; Shi et al. 1998; Zou et al. 1998). This chaperone-HSF1 equilibrium is disrupted presumably by the rapid appearance of non-native species that redirect chaperones to sequester these non-native species, thus releasing HSF1 to self-associate and form trimers.

The HSP70 and HSP90 molecular chaperones have been identified by genetic and biochemical studies as major regulators of HSF1 activity (Craig and Gross 1991; Abravaya et al. 1992; Baler et al. 1992, 1996; Mosser et al. 1993; Ali et al. 1998; Shi et al. 1998; Zou et al. 1998). Consequently, depletion or over-expression of these chaperones results in the activation or repression, respectively, of HSF1, suggesting a direct regulatory feedback loop. Regulation of HSF1 by chaperones provides a satisfying model for the HSR, whereby under control, non-stress physiological states, the chaperones HSP70 and HSP90 are sufficiently abundant to bind and repress HSF1. Upon heat shock, misfolded proteins accumulate rapidly, thus titrating HSP70 and HSP90 away from HSF1 and towards non-native proteins, leading to induction of the HSR to rebalance proteostasis. This chaperone titration model seems to function not only in eukaryotes but in at least two distinct types of HSRs present in prokaryotes (Guisbert et al. 2008). An additional, untested hypothesis is that a similar chaperone titration model may help to explain how blocks to protein synthesis and splicing are regulated.

In addition to chaperones, other regulators of the HSR have been identified. For example, HDAC6 mediates deacetylation of HSP90, enhancing its activity and association with its substrates including HSF1 (Bali et al. 2005; Kovacs et al. 2005; Boyault et al. 2007). Additionally, HSF2, an orthologue of HSF1, is activated by heat shock and can form hetero-oligomers with HSF1 (Ostling et al. 2007; Sandqvist et al. 2009). Because HSF2 is principally expressed during development, this could lead to activation of HSF1. To date, however, there has not been a comprehensive genetic analysis of the HSR regulatory pathway, in part because traditional forward genetic screens are inadequately suited to identify genes that regulate the HSR. These approaches depend on the introduction of mutations that can destabilize the folding of the target protein, resulting in expression of misfolded species and leading to indirect induction of the HSR. Indeed, a forward genetic screen in *Drosophila* for regulators of the HSR identified such mutations that destabilized a muscle-specific actin and lead to induction of the HSR (Bonner et al. 1984; Parker-Thornburg and Bonner 1987).

Systems biology has emerged as a powerful new approach to understanding complex pathways. A key component of this approach is the development of mathematical models that can recapitulate the features of a biological response. Several mathematical models involving the HSR have been developed in eukaryotes (Rieger et al. 2005; Lipan et al. 2007; Dwivedi et al. 2008; Petre et al. 2011). Two of these models focus on HSR regulation and recapitulate many of the dynamics of the response, but they do not accurately predict the dynamics of HSR attenuation (Rieger et al. 2005; Petre 2011). Importantly, these models do not include some of the more recent advances in HSR regulation, such as SUMOylation and acetylation; however, it is also likely that there are additional, uncharacterized regulatory mechanisms.

The HSR in multicellular organisms

A recent observation on the regulation of the HSR in *C. elegans* revealed additional layers of regulatory complexity. First, it was shown that neuronal signaling can affect induction of the HSR in the entire organism, reflecting non-cell autonomous regulation (Prahlad et al. 2008). Second, there are profound differences in HSR regulation in different cell types (Bienz 1984; Sprang and Brown 1987; Marini et al. 1990; Nishimura et al. 1991; Pardue et al. 1992; Mathur et al. 1994; Marcuccilli et al. 1996; Voisin et al. 1996; Batulan et al. 2003). For example, some retinoblastoma cells induce HSP90 but not HSP70, whereas other cells induce both (Mathur et al. 1994). Also, there are differences between soma and germline in the developing drosophila and between different tissues in rats exposed to hyperthermia (Blake et al. 1990; Wang and Lindquist 1998). These observations demonstrate that the extent and nature of the HSR can be modified in different cell and tissue types, yet the precise mechanisms and results of this regulation remain to be elucidated.

Changes to the regulation of the HSR are also observed during the course of development. Conditional and null alleles of HSF1 have specific defects during development as well as impaired survival at elevated temperatures (Sorger and Pelham 1988; Jedlicka et al. 1997; Xiao et al. 1999; Christians et al. 2000). Further, some organisms are unable to mount a complete HSR until they reach a specific developmental stage (Bienz 1984; Wang and Lindquist 1998). For example, in *Drosophila*, nuclear transport of HSF1 is mediated by karyopherin-a3, which is not expressed in the early embryo (Fang et al. 2001). However, the roles and regulation of HSF1 in development remain largely unexplored.

The HSR in aging and disease

Protein aggregation and the collateral damage associated with the stress of misfolded proteins are fundamental problems in the biology of aging and human disease, raising the question why the HSR and other stress responses that are so essential and effective to protect cells against acute stress are seemingly less effective in protecting tissues in diseases of protein conformation (Fig. 2). Presumably this is not just a matter of chronic expression of damaged proteins, as the biological noise emanating from intrinsic error in protein biosynthesis, genetic mutations, and polymorphisms is itself quite substantial (Drummond and Wilke 2008) The accumulation of aggregates during aging and the age-associated disruption in proteostasis lead to an amplification of protein damage that contributes to cellular toxicity (Stadtman 1992; Morley et al. 2002; Hsu et al. 2003; Cohen et al. 2006).

The relationship between life span and cumulative protein damage is provocative and supported by evidence that life span regulation by the insulin-like signaling (ILS) pathway and insulin growth factor 1 receptor (IGF1R) is dependent upon HSF1 (Hsu et al. 2003; Morley and Morimoto 2004). Genetic knockdown of HSF1 increases aggregation and leads to premature aging, in addition to suppressing life span-enhancing pathways. HSF1 activity, in response to heat shock, declines precipitously in early adulthood of *C. elegans,* and is associated with the loss-of-function of proteins with metastable, temperature-sensitive phenotypes. Consistent with these observations, life span-enhancing mutations in the ILS pathway suppress aggregation and cellular dysfunction in *C. elegans* models of Aβ, polyQ, and ataxin-3 (Morley et al. 2002; Cohen et al. 2006).

Another line of support linking HSF1 activity with cellular pathways implicated in metabolism and life span is the role of the NAD-dependent sirtruin, SiRT1, in the regulation of the HSR (Morley et al. 2002; Hsu et al. 2003; Lee et al. 2003; Brunet et al. 2004; Morley and Morimoto 2004; Westerheide et al. 2009). SiRT1 deacetylation in the DNA binding domain of HSF1 maintains HSF1 functional trimers in a DNA binding-competent state bound to the HSE and engaged for transcription of HS genes. Regulatory override can be achieved by enhancing

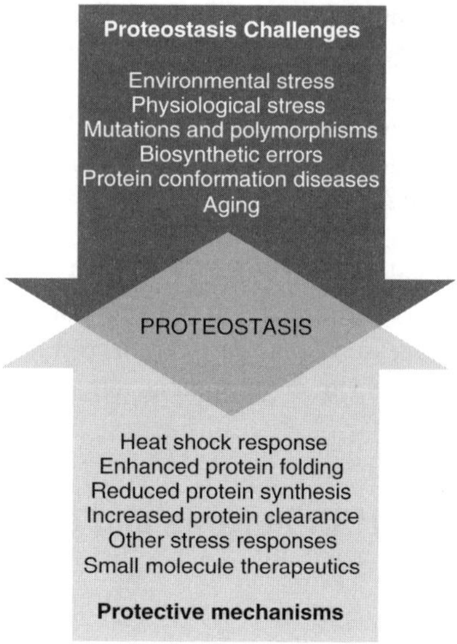

Fig. 2 Protective mechanisms help to maintain proteostasis in the face of constant challenges. Proteostasis challenges of stress, aging, and disease are balanced by adaptive and molecular and cellular protective responses

HSF1 by altering the levels of NAD, therefore prolonging the "youthfulness" of the proteostatic machineries to ensure that the regulation of proteostasis remains robust.

The relationship between proteostasis decline and aging is a striking feature common to all protein conformational diseases. For neurodegenerative disease, the age at which neurological symptoms appear varies, with Alzheimer's disease and Parkinson's disease being late onset, Huntington's disease showing symptoms during mid-to-late-life, and ALS occurring in early to mid-life. For each of these, as with other conformational diseases, a striking characteristic is that the associated mutant protein shares no similarity in primary sequence, activity, or function but does share common biophysical features of conformational instability and aggregation (Chiti and Dobson 2006). The transition from soluble to aggregated states is exemplified by the triplet-repeat (CAG) diseases of proteins with expanded polyQ proteins that result in Huntington's disease, spinocerebellar ataxias, and Kennedy's disease (Perutz 1999; Zoghbi and Orr 2000; Ross 2002). In *C. elegans* and Drosophila models, the expression of polyQ repeats of different lengths in neurons, intestine, and muscle cells has revealed a clear relationship between polyQ length, aggregation, and toxicity (Warrick et al. 1998; Faber et al. 1999; Marsh et al. 2000; Satyal et al. 2000; Parker et al. 2001; Morley et al. 2002; Takeyama et al. 2002; Teixeira-Castro et al. 2011). In young adult animals expressing polyQ in body wall muscle cells and neurons, $<$Q35 is diffuse and soluble whereas $>$ Q40 forms insoluble aggregates (Satyal et al. 2000; Morley et al. 2002). Upon aging, Q35 animals show age-dependent aggregation and toxicity. The appearance of polyQ aggregates in neurons and muscle cells is associated with toxicity that can be suppressed by downregulation of the ILS pathway, activation of HSF1, and elevated expression of chaperones and other components of the proteostasis machineries (Morley et al. 2002; Brignull et al. 2006).

The aggregation and toxicity phenotypes of polyQ in *C. elegans* have been invaluable to probe the genetics of proteostasis. Genome-wide RNAi screens of *C. elegans* expressing Q35-YFP have identified ~340 genetic modifiers that either enhance or inhibit the PN and thus increase or decrease polyQ aggregation (Nollen et al. 2004). These modifiers cluster into five major pathways encoding genes involved in RNA metabolism, protein synthesis, protein folding, protein trafficking and protein degradation. Downregulation of the RNA-processing machinery by RNAi accelerates polyQ aggregation, perhaps due to an increased burden or imbalance of abnormal proteins. Likewise, enhancing cholinergic signaling at the neuromuscular junction induces the Ca^{2+}-dependent activation of HSF1 and expression of cytoplasmic chaperones (Silva, Amaral, and Morimoto, unpublished observations), resulting in a reprogramming of body wall muscle cells by neuronal signaling to have an enhanced state of proteostasis that can suppress polyQ aggregation and maintain temperature-sensitvie (TS) proteins in a soluble and functional state. What appears to be important is to achieve a precise balance of neuronal signaling; the complete loss of the inhibitory GABAergic signal at the neuromuscular junction results in hyperstimulation of muscle cells and enhanced aggregation (Garcia et al. 2007). Together, these results reveal that protein homeostasis is strongly dependent upon cell non-autonomous regulation by the nervous system.

We propose that the susceptibility of the cellular proteostasis environment to conformational disease is a direct consequence of multiple events initiated by age-dependent changes in the PN and amplified by cumulative protein damage. In addition to the decline of the HSR and unfolded protein response (UPR) in early adulthood in *C. elegans*, the expression of disease-associated, aggregation-prone proteins has global consequences on the overall stability of the proteome. For example, metastable proteins harboring missense mutations that result in TS phenotypes become targets for aggregation when co-expressed with expanded polyQ proteins or mutant SOD1 in neurons and muscle cells (Gidalevitz et al. 2006, 2009, 2010), revealing that the proteome expressed in these tissues does not have additional capacity for non-native states and that the PN is delicately balanced. The age-associated decline in the PN is likely due to a compromise in the robustness of cell stress responses and a reduced capacity of chaperones and clearance machines (Stadtman 1992; Shamovsky and Gershon 2004; Ben-Zvi et al. 2009).

Proteostasis Therapeutics and the Regulation of HSF1 by Small Molecules

Regulation of the HSR by small molecules (Fig. 3) not only provides chemical genetic tools for the analysis of the HSR but also establishes a therapeutic strategy to enhance the cellular proteostasis environment to prevent misfolding of diverse disease-associated proteins (Westerheide and Morimoto 2005; Calamini et al. 2012). This strategy targeted at the PN and the cellular environment is an alternative to correcting folding instability by small molecule pharmacological chaperones.

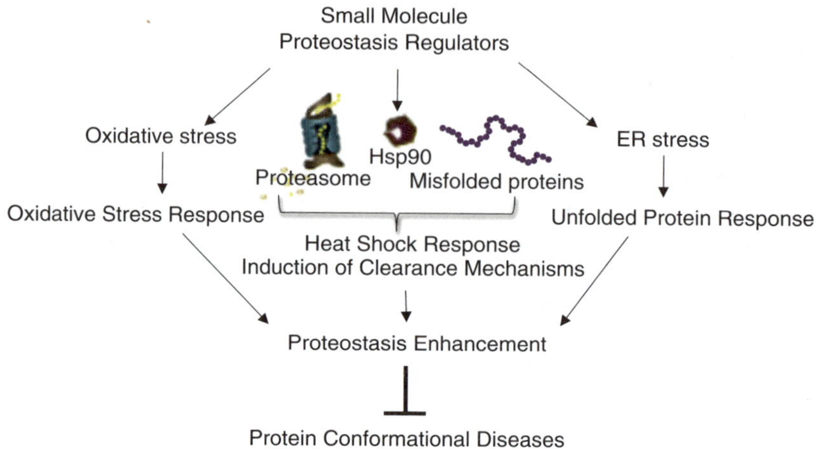

Fig. 3 Strategies for small molecule proteostasis regulators for diseases of protein conformation. Targets for small molecules that activate the primary components of the proteostasis network include the heat shock response, unfolded protein response, oxidative stress response, and clearance mechanisms (ubiquitin proteasome and autophagy). The resulting proteostasis enhancement then prevents the further accumulation of damaged proteins that occurs in diseases of protein conformation

A number of compounds with diverse activities that enhance the PN include protease inhibitors (DCIC, TPCK, and TLCK) and proteasome inhibitors (MG132, lactacystin) that induce the HSR by increasing the pool of non-native proteins targeted to the proteasome (Mathew et al. 1998; Rossi et al. 1998; Holmberg et al. 2000), and the Hsp90 inhibitors (radicicol, geldanamycin, and 17AAG) that activate the HSR (Bagatell et al. 2000; Whitesell et al. 2003, Bagatell and Whitesell 2004; Whitesell and Lindquist 2005; Sharp and Workman 2006). Other pharmacologically active small molecules that induce the HSR include the NSAID, sodium salicylate, which activates HSF1 DNA binding and synergizes with other stress conditions to induce a complete HSR (Jurivich et al. 1992). Other molecules associated with inflammation, including the widely used NSAID, Indomethacin, and inflammatory modulators such as arachidonic acid, and the cyclopentenone prostaglandins (PGA1, PGA2 and PGJ2), all induce HSF1 DNA binding and the HSR (Ohno et al. 1988; Amici et al. 1992; Jurivich et al. 1994; Lee et al. 1995). Among the more widely used small molecules for induction of the HSR is the triterpenoid celastrol isolated from the Chinese plant *Triptergium wilfordii,* which induces multiple stress responses including the HSR, UPR, and the anti-oxidant response (Westerheide et al. 2004; Trott et al. 2008).

A recent large-scale (~1million), small molecule screen performed in human tissue culture cells for HSF1-dependent proteostasis regulators (PRs) identified ~300 chemical inducers of the HSR (Calamini et al. 2012). These PRs were classified by their stress response signatures as: HSR and HSF1-selective, induction of both HSF1 and the lumen UPR, and activation of the HSF1, UPR, and the anti-oxidant response (Fig. 3). These properties, to activate multiple stress response pathways, were used to complement the cell's biological response to specific aggregation-prone proteins and to correct the proteostatic deficiency. For example, small molecule inducers of the HSR that restored the cytoplasmic PN were effective in the suppression of polyQ-mediated proteotoxicity but insufficient to correct the folding of mutant CFTR. Folding of mutant CFTR, however, was enhanced when the UPR was co-induced with the HSR, and when the anti-oxidant response was co-induced with HSR, and UPR. Oxidative stress is associated with the pathogenesis of many diseases, including Alzheimer's disease, Parkinson's disease, amyotropic lateral sclerosis, and cystic fibrosis; thus activation of the anti-oxidant response pathway in conjunction with the HSR may have broad beneficial implications. In support of this hypothesis, PRs that induced multiple stress-protective pathways were highly effective in multiple disease models affecting distinct cellular compartments, which reveals that the redundant and overlapping profiles of PRs, with HSF1 at the stress network hub, function in concert with other stress-signaling pathways to coordinate the PN. By this approach, the growing knowledge of the PN and stress biology can be used to promote the health of the cell by tuning the efficacy of small molecule PRs by their stress response signatures to manage the imbalance caused by diverse misfolding-prone proteins. The strategy of PN adjustment by small molecule PRs provides a previously undeveloped approach to restore cellular proteostatic capabilities and thus achieve proteome balance in both loss- and gain-of-function diseases.

Summary

The HSR is a universal, highly conserved stress response to increased temperature and cytoplasmic protein misfolding. This response is critical not only for cellular homeostasis but also for organismal development and aging. Further, it is associated with numerous major human diseases, including cancer and neurodegeneration. Many details of this extremely complicated response have been uncovered, yet despite its importance, many central questions remain to be answered to provide a fully integrated picture of HSR regulation.

References

Abravaya K, Phillips B, Morimoto RI (1991) Heat shock-induced interactions of heat shock transcription factor and the human hsp70 promoter examined by in vivo footprinting. Mol Cell Biol 11:586–592

Abravaya K, Myers MP, Murphy SP, Morimoto RI (1992) The human heat shock protein hsp70 interacts with HSF, the transcription factor that regulates heat shock gene expression. Genes Dev 6:1153–1164

Akerfelt M, Morimoto RI, Sistonen L (2010) Heat shock factors: integrators of cell stress, development and lifespan. Nat Rev Mol Cell Biol 11:545–555

Ali A, Bharadwaj S, O'Carroll R, Ovsenek N (1998) HSP90 interacts with and regulates the activity of heat shock factor 1 in Xenopus oocytes. Mol Cell Biol 18:4949–4960

Allen TA, Von Kaenel S, Goodrich JA, Kugel JF (2004) The SINE-encoded mouse B2 RNA represses mRNA transcription in response to heat shock. Nat Struct Mol Biol 11:816–821

Amici C, Sistonen L, Santoro MG, Morimoto RI (1992) Antiproliferative prostaglandins activate heat shock transcription factor. Proc Natl Acad Sci USA 89:6227–6231

Anckar J, Sistonen L (2007) Heat shock factor 1 as a coordinator of stress and developmental pathways. Adv Exp Med Biol 594:78–88

Anckar J, Hietakangas V, Denessiouk K, Thiele DJ, Johnson MS, Sistonen L (2006) Inhibition of DNA binding by differential sumoylation of heat shock factors. Mol Cell Biol 26:955–964

Bagatell R, Whitesell L (2004) Altered Hsp90 function in cancer: a unique therapeutic opportunity. Mol Cancer Ther 3:1021–1030

Bagatell R, Paine-Murrieta GD, Taylor CW, Pulcini EJ, Akinaga S, Benjamin IJ, Whitesell L (2000) Induction of a heat shock factor 1-dependent stress response alters the cytotoxic activity of hsp90-binding agents. Clin Cancer Res 6:3312–3318

Baler R, Welch WJ, Voellmy R (1992) Heat shock gene regulation by nascent polypeptides and denatured proteins: hsp70 as a potential autoregulatory factor. J Cell Biol 117:1151–1159

Baler R, Zou J, Voellmy R (1996) Evidence for a role of Hsp70 in the regulation of the heat shock response in mammalian cells. Cell Stress Chaperones 1:33–39

Bali P, Pranpat M, Bradner J, Balasis M, Fiskus W, Guo F, Rocha K, Kumaraswamy S, Boyapalle S, Atadja P, Seto E, Bhalla K (2005) Inhibition of histone deacetylase 6 acetylates and disrupts the chaperone function of heat shock protein 90: a novel basis for antileukemia activity of histone deacetylase inhibitors. J Biol Chem 280:26729–26734

Batulan Z, Shinder GA, Minotti S, He BP, Doroudchi MM, Nalbantoglu J, Strong MJ, Durham HD (2003) High threshold for induction of the stress response in motor neurons is associated with failure to activate HSF1. J Neurosci 23:5789–5798

Ben-Zvi A, Miller EA, Morimoto RI (2009) Collapse of proteostasis represents an early molecular event in Caenorhabditis elegans aging. Proc Natl Acad Sci USA 106:14914–14919

Bienz M (1984) Developmental control of the heat shock response in Xenopus. Proc Natl Acad Sci USA 81:3138–3142

Bijur GN, Jope RS (2000) Opposing actions of phosphatidylinositol 3-kinase and glycogen synthase kinase-3beta in the regulation of HSF-1 activity. J Neurochem 75:2401–2408

Birch-Machin I, Gao S, Huen D, McGirr R, White RA, Russell S (2005) Genomic analysis of heat-shock factor targets in Drosophila. Genome Biol 6:R63

Blake MJ, Gershon D, Fargnoli J, Holbrook NJ (1990) Discordant expression of heat shock protein mRNAs in tissues of heat-stressed rats. J Biol Chem 265:15275–15279

Boehm AK, Saunders A, Werner J, Lis JT (2003) Transcription factor and polymerase recruitment, modification, and movement on dhsp70 in vivo in the minutes following heat shock. Mol Cell Biol 23:7628–7637

Bonanou-Tzedaki SA, Smith KE, Sheeran BA, Arnstein HR (1978) The high-temperature inactivation of rabbit reticulocyte lysates by a haemin-independent mechanism. Eur J Biochem 84:591–600

Bonner JJ, Parks C, Parker-Thornburg J, Mortin MA, Pelham HR (1984) The use of promoter fusions in Drosophila genetics: isolation of mutations affecting the heat shock response. Cell 37:979–991

Boyault C, Zhang Y, Fritah S, Caron C, Gilquin B, Kwon SH, Garrido C, Yao TP, Vourc'h C, Matthias P, Khochbin S (2007) HDAC6 controls major cell response pathways to cytotoxic accumulation of protein aggregates. Genes Dev 21:2172–2181

Brignull HR, Moore FE, Tang SJ, Morimoto RI (2006) Polyglutamine proteins at the pathogenic threshold display neuron-specific aggregation in a pan-neuronal Caenorhabditis elegans model. J Neurosci 26:7597–7606

Brunet A, Sweeney LB, Sturgill JF, Chua KF, Greer PL, Lin Y, Tran H, Ross SE, Mostoslavsky R, Cohen HY, Hu LS, Cheng HL, Jedrychowski MP, Gygi SP, Sinclair DA, Alt FW, Greenberg

ME (2004) Stress-dependent regulation of FOXO transcription factors by the SIRT1 deacetylase. Science 303:2011–2015

Calamini B, Silva MC, Madoux F, Hutt DM, Khanna S, Chalfant MA, Saldanha SA, Hodder P, Tait BD, Garza D, Balch WE, Morimoto RI (2012) Small-molecule proteostasis regulators for protein conformational diseases. Nat Chem Biol 8:185–196

Chiti F, Dobson CM (2006) Protein misfolding, functional amyloid, and human disease. Annu Rev Biochem 75:333–366

Christians E, Davis AA, Thomas SD, Benjamin IJ (2000) Maternal effect of Hsf1 on reproductive success. Nature 407:693–694

Chu B, Soncin F, Price BD, Stevenson MA, Calderwood SK (1996) Sequential phosphorylation by mitogen-activated protein kinase and glycogen synthase kinase 3 represses transcriptional activation by heat shock factor-1. J Biol Chem 271:30847–30857

Chu B, Zhong R, Soncin F, Stevenson MA, Calderwood SK (1998) Transcriptional activity of heat shock factor 1 at 37 degrees C is repressed through phosphorylation on two distinct serine residues by glycogen synthase kinase 3 and protein kinases Calpha and Czeta. J Biol Chem 273:18640–18646

Cohen E, Bieschke J, Perciavalle RM, Kelly JW, Dillin A (2006) Opposing activities protect against age-onset proteotoxicity. Science 313:1604–1610

Cotto JJ, Kline M, Morimoto RI (1996) Activation of heat shock factor 1 DNA binding precedes stress-induced serine phosphorylation. Evidence for a multistep pathway of regulation. J Biol Chem 271:3355–3358

Craig EA, Gross CA (1991) Is hsp70 the cellular thermometer? Trends Biochem Sci 16:135–140

Dai R, Frejtag W, He B, Zhang Y, Mivechi NF (2000) c-Jun NH2-terminal kinase targeting and phosphorylation of heat shock factor-1 suppress its transcriptional activity. J Biol Chem 275:18210–18218

Denegri M, Moralli D, Rocchi M, Biggiogera M, Raimondi E, Cobianchi F, De Carli L, Riva S, Biamonti G (2002) Human chromosomes 9, 12, and 15 contain the nucleation sites of stress-induced nuclear bodies. Mol Biol Cell 13:2069–2079

Drummond DA, Wilke CO (2008) Mistranslation-induced protein misfolding as a dominant constraint on coding-sequence evolution. Cell 134:341–352

Dwivedi A, Karan BM, Das BN, Sinha RK (2008) Digital-analog hybrid control model for eukaryotic heat shock response illustrating the dynamics of heat shock protein 70 on exposure to thermal stress. Comput Methods Programs Biomed 90:17–24

Faber PW, Alter JR, MacDonald ME, Hart AC (1999) Polyglutamine-mediated dysfunction and apoptotic death of a Caenorhabditis elegans sensory neuron. Proc Natl Acad Sci USA 96:179–184

Fang X, Chen T, Tran K, Parker CS (2001) Developmental regulation of the heat shock response by nuclear transport factor karyopherin-alpha3. Development 128:3349–3358

Garcia SM, Casanueva MO, Silva MC, Amaral MD, Morimoto RI (2007) Neuronal signaling modulates protein homeostasis in Caenorhabditis elegans post-synaptic muscle cells. Genes Dev 21:3006–3016

Gidalevitz T, Ben-Zvi A, Ho KH, Brignull HR, Morimoto RI (2006) Progressive disruption of cellular protein folding in models of polyglutamine diseases. Science 311:1471–1474

Gidalevitz T, Krupinski T, Garcia S, Morimoto RI (2009) Destabilizing protein polymorphisms in the genetic background direct phenotypic expression of mutant SOD1 toxicity. PLoS Genet 5: e1000399

Gidalevitz T, Kikis EA, Morimoto RI (2010) A cellular perspective on conformational disease: the role of genetic background and proteostasis networks. Curr Opin Struct Biol 20:23–32

Greene JM, Larin Z, Taylor IC, Prentice H, Gwinn KA, Kingston RE (1987) Multiple basal elements of a human hsp70 promoter function differently in human and rodent cell lines. Mol Cell Biol 7:3646–3655

Guertin MJ, Lis JT (2010) Chromatin landscape dictates HSF binding to target DNA elements. PLoS Genet 6

Guettouche T, Boellmann F, Lane WS, Voellmy R (2005) Analysis of phosphorylation of human heat shock factor 1 in cells experiencing a stress. BMC Biochem 6:4

GuhaThakurta D, Palomar L, Stormo GD, Tedesco P, Johnson TE, Walker DW, Lithgow G, Kim S, Link CD (2002) Identification of a novel cis-regulatory element involved in the heat shock response in Caenorhabditis elegans using microarray gene expression and computational methods. Genome Res 12:701–712

Guisbert E, Yura T, Rhodius VA, Gross CA (2008) Convergence of molecular, modeling, and systems approaches for an understanding of the Escherichia coli heat shock response. Microbiol Mol Biol Rev 72:545–554

Hahn JS, Hu Z, Thiele DJ, Iyer VR (2004) Genome-wide analysis of the biology of stress responses through heat shock transcription factor. Mol Cell Biol 24:5249–5256

Hajdu-Cronin YM, Chen WJ, Sternberg PW (2004) The L-type cyclin CYL-1 and the heat-shock-factor HSF-1 are required for heat-shock-induced protein expression in Caenorhabditis elegans. Genetics 168:1937–1949

He B, Meng YH, Mivechi NF (1998) Glycogen synthase kinase 3beta and extracellular signal-regulated kinase inactivate heat shock transcription factor 1 by facilitating the disappearance of transcriptionally active granules after heat shock. Mol Cell Biol 18:6624–6633

Hietakangas V, Ahlskog JK, Jakobsson AM, Hellesuo M, Sahlberg NM, Holmberg CI, Mikhailov A, Palvimo JJ, Pirkkala L, Sistonen L (2003) Phosphorylation of serine 303 is a prerequisite for the stress-inducible SUMO modification of heat shock factor 1. Mol Cell Biol 23:2953–2968

Holmberg CI, Illman SA, Kallio M, Mikhailov A, Sistonen L (2000) Formation of nuclear HSF1 granules varies depending on stress stimuli. Cell Stress Chaperones 5:219–228

Holmberg CI, Hietakangas V, Mikhailov A, Rantanen JO, Kallio M, Meinander A, Hellman J, Morrice N, MacKintosh C, Morimoto RI, Eriksson JE, Sistonen L (2001) Phosphorylation of serine 230 promotes inducible transcriptional activity of heat shock factor 1. EMBO J 20:3800–3810

Hsu AL, Murphy CT, Kenyon C (2003) Regulation of aging and age-related disease by DAF-16 and heat-shock factor. Science 300:1142–1145

Hurst R, Matts RL (1987) Phosphorothioated binary complex of eukaryotic initiation factor eIF-2 and GDP inhibits protein synthesis in hemin-supplemented reticulocyte lysates. Biochem Biophys Res Commun 147:772–777

Jedlicka P, Mortin MA, Wu C (1997) Multiple functions of Drosophila heat shock transcription factor in vivo. EMBO J 16:2452–2462

Jolly C, Konecny L, Grady DL, Kutskova YA, Cotto JJ, Morimoto RI, Vourc'h C (2002) In vivo binding of active heat shock transcription factor 1 to human chromosome 9 heterochromatin during stress. J Cell Biol 156:775–781

Jolly C, Metz A, Govin J, Vigneron M, Turner BM, Khochbin S, Vourc'h C (2004) Stress-induced transcription of satellite III repeats. J Cell Biol 164:25–33

Jurivich DA, Sistonen L, Kroes RA, Morimoto RI (1992) Effect of sodium salicylate on the human heat shock response. Science 255:1243–1245

Jurivich DA, Sistonen L, Sarge KD, Morimoto RI (1994) Arachidonate is a potent modulator of human heat shock gene transcription. Proc Natl Acad Sci U S A 91:2280–2284

Kedersha NL, Gupta M, Li W, Miller I, Anderson P (1999) RNA-binding proteins TIA-1 and TIAR link the phosphorylation of eIF-2 alpha to the assembly of mammalian stress granules. J Cell Biol 147:1431–1442

Kim J, Nueda A, Meng YH, Dynan WS, Mivechi NF (1997) Analysis of the phosphorylation of human heat shock transcription factor-1 by MAP kinase family members. J Cell Biochem 67:43–54

Kline MP, Morimoto RI (1997) Repression of the heat shock factor 1 transcriptional activation domain is modulated by constitutive phosphorylation. Mol Cell Biol 17:2107–2115

Knauf U, Newton EM, Kyriakis J, Kingston RE (1996) Repression of human heat shock factor 1 activity at control temperature by phosphorylation. Genes Dev 10:2782–2793

Kovacs JJ, Murphy PJ, Gaillard S, Zhao X, Wu JT, Nicchitta CV, Yoshida M, Toft DO, Pratt WB, Yao TP (2005) HDAC6 regulates Hsp90 acetylation and chaperone-dependent activation of glucocorticoid receptor. Mol Cell 18:601–607

Larson JS, Schuetz TJ, Kingston RE (1988) Activation in vitro of sequence-specific DNA binding by a human regulatory factor. Nature 335:372–375

Lee BS, Chen J, Angelidis C, Jurivich DA, Morimoto RI (1995) Pharmacological modulation of heat shock factor 1 by antiinflammatory drugs results in protection against stress-induced cellular damage. Proc Natl Acad Sci U S A 92:7207–7211

Lee SS, Kennedy S, Tolonen AC, Ruvkun G (2003) DAF-16 target genes that control C. elegans life-span and metabolism. Science 300:644–647

Lipan O, Navenot JM, Wang Z, Huang L, Peiper SC (2007) Heat shock response in CHO mammalian cells is controlled by a nonlinear stochastic process. PLoS Comput Biol 3:1859–1870

Lis JT, Mason P, Peng J, Price DH, Werner J (2000) P-TEFb kinase recruitment and function at heat shock loci. Genes Dev 14:792–803

Lu Q, Wallrath LL, Granok H, Elgin SC (1993) (CT)n (GA)n repeats and heat shock elements have distinct roles in chromatin structure and transcriptional activation of the Drosophila hsp26 gene. Mol Cell Biol 13:2802–2814

Marcuccilli CJ, Mathur SK, Morimoto RI, Miller RJ (1996) Regulatory differences in the stress response of hippocampal neurons and glial cells after heat shock. J Neurosci 16:478–485

Mariner PD, Walters RD, Espinoza CA, Drullinger LF, Wagner SD, Kugel JF, Goodrich JA (2008) Human Alu RNA is a modular transacting repressor of mRNA transcription during heat shock. Mol Cell 29:499–509

Marini AM, Kozuka M, Lipsky RH, Nowak TS Jr (1990) 70-kilodalton heat shock protein induction in cerebellar astrocytes and cerebellar granule cells in vitro: comparison with immunocytochemical localization after hyperthermia in vivo. J Neurochem 54:1509–1516

Marsh JL, Walker H, Theisen H, Zhu YZ, Fielder T, Purcell J, Thompson LM (2000) Expanded polyglutamine peptides alone are intrinsically cytotoxic and cause neurodegeneration in Drosophila. Hum Mol Genet 9:13–25

Mathew A, Mathur SK, Morimoto RI (1998) Heat shock response and protein degradation: regulation of HSF2 by the ubiquitin-proteasome pathway. Mol Cell Biol 18:5091–5098

Mathur SK, Sistonen L, Brown IR, Murphy SP, Sarge KD, Morimoto RI (1994) Deficient induction of human hsp70 heat shock gene transcription in Y79 retinoblastoma cells despite activation of heat shock factor 1. Proc Natl Acad Sci U S A 91:8695–8699

Matts RL, Hurst R (1989) Evidence for the association of the heme-regulated eIF-2 alpha kinase with the 90-kDa heat shock protein in rabbit reticulocyte lysate in situ. J Biol Chem 264:15542–15547

Matts RL, Levin DH, London IM (1983) Effect of phosphorylation of the alpha-subunit of eukaryotic initiation factor 2 on the function of reversing factor in the initiation of protein synthesis. Proc Natl Acad Sci U S A 80:2559–2563

McMillan DR, Xiao X, Shao L, Graves K, Benjamin IJ (1998) Targeted disruption of heat shock transcription factor 1 abolishes thermotolerance and protection against heat-inducible apoptosis. J Biol Chem 273:7523–7528

Mendez R, de Haro C (1994) Casein kinase II is implicated in the regulation of heme-controlled translational inhibitor of reticulocyte lysates. J Biol Chem 269:6170–6176

Mendez R, Moreno A, de Haro C (1992) Regulation of heme-controlled eukaryotic polypeptide chain initiation factor 2 alpha-subunit kinase of reticulocyte lysates. J Biol Chem 267:11500–11507

Morgan WD, Williams GT, Morimoto RI, Greene J, Kingston RE, Tjian R (1987) Two transcriptional activators, CCAAT-box-binding transcription factor and heat shock transcription factor, interact with a human hsp70 gene promoter. Mol Cell Biol 7:1129–1138

Morimoto RI (1998) Regulation of the heat shock transcriptional response: cross talk between a family of heat shock factors, molecular chaperones, and negative regulators. Genes Dev 12:3788–3796

Morimoto RI (2008) Proteotoxic stress and inducible chaperone networks in neurodegenerative disease and aging. Genes Dev 22:1427–1438

Morimoto RI (2011) The heat shock response: systems biology of proteotoxic stress in aging and disease. Cold Spring Harb Symp Quant Biol 76:91–99

Morley JF, Morimoto RI (2004) Regulation of longevity in Caenorhabditis elegans by heat shock factor and molecular chaperones. Mol Biol Cell 15:657–664

Morley JF, Brignull HR, Weyers JJ, Morimoto RI (2002) The threshold for polyglutamine-expansion protein aggregation and cellular toxicity is dynamic and influenced by aging in Caenorhabditis elegans. Proc Natl Acad Sci U S A 99:10417–10422

Mosser DD, Duchaine J, Massie B (1993) The DNA-binding activity of the human heat shock transcription factor is regulated in vivo by hsp70. Mol Cell Biol 13:5427–5438

Nishimura RN, Dwyer BE, Clegg K, Cole R, de Vellis J (1991) Comparison of the heat shock response in cultured cortical neurons and astrocytes. Brain Res Mol Brain Res 9:39–45

Nollen EA, Garcia SM, van Haaften G, Kim S, Chavez A, Morimoto RI, Plasterk RH (2004) Genome-wide RNA interference screen identifies previously undescribed regulators of polyglutamine aggregation. Proc Natl Acad Sci U S A 101:6403–6408

Nonaka G, Blankschien M, Herman C, Gross CA, Rhodius VA (2006) Regulon and promoter analysis of the E. coli heat-shock factor, sigma32, reveals a multifaceted cellular response to heat stress. Genes Dev 20:1776–1789

Ohno K, Fukushima M, Fujiwara M, Narumiya S (1988) Induction of 68,000-dalton heat shock proteins by cyclopentenone prostaglandins. Its association with prostaglandin-induced G1 block in cell cycle progression. J Biol Chem 263:19764–19770

Ostling P, Bjork JK, Roos-Mattjus P, Mezger V, Sistonen L (2007) Heat shock factor 2 (HSF2) contributes to inducible expression of hsp genes through interplay with HSF1. J Biol Chem 282:7077–7086

Page TJ, Sikder D, Yang L, Pluta L, Wolfinger RD, Kodadek T, Thomas RS (2006) Genome-wide analysis of human HSF1 signaling reveals a transcriptional program linked to cellular adaptation and survival. Mol Biosyst 2:627–639

Pardue S, Groshan K, Raese JD, Morrison-Bogorad M (1992) Hsp70 mRNA induction is reduced in neurons of aged rat hippocampus after thermal stress. Neurobiol Aging 13:661–672

Parker JA, Connolly JB, Wellington C, Hayden M, Dausset J, Neri C (2001) Expanded polyglutamines in Caenorhabditis elegans cause axonal abnormalities and severe dysfunction of PLM mechanosensory neurons without cell death. Proc Natl Acad Sci U S A 98:13318–13323

Parker-Thornburg J, Bonner JJ (1987) Mutations that induce the heat shock response of Drosophila. Cell 51:763–772

Perutz MF (1999) Glutamine repeats and neurodegenerative diseases: molecular aspects. Trends Biochem Sci 24:58–63

Petesch SJ, Lis JT (2008) Rapid, transcription-independent loss of nucleosomes over a large chromatin domain at Hsp70 loci. Cell 134:74–84

Petre I, Hyder CL, Mizera A, Meinander A, Mikhailov A, Morimoto RI, Sistonen L, Eriksson JE, Back R-J (2011) A simple mass-action model for the eukaryotic heat shock response and its mathematical validation. Natural Computing 10:595–612

Pirkkala L, Nykanen P, Sistonen L (2001) Roles of the heat shock transcription factors in regulation of the heat shock response and beyond. FASEB J 15:1118–1131

Prahlad V, Cornelius T, Morimoto RI (2008) Regulation of the cellular heat shock response in Caenorhabditis elegans by thermosensory neurons. Science 320:811–814

Rieger TR, Morimoto RI, Hatzimanikatis V (2005) Mathematical modeling of the eukaryotic heat-shock response: dynamics of the hsp70 promoter. Biophys J 88:1646–1658

Rizzi N, Denegri M, Chiodi I, Corioni M, Valgardsdottir R, Cobianchi F, Riva S, Biamonti G (2004) Transcriptional activation of a constitutive heterochromatic domain of the human genome in response to heat shock. Mol Biol Cell 15:543–551

Ross CA (2002) Polyglutamine pathogenesis: emergence of unifying mechanisms for Huntington's disease and related disorders. Neuron 35:819–822

Rossi A, Elia G, Santoro MG (1998) Activation of the heat shock factor 1 by serine protease inhibitors. An effect associated with nuclear factor-kappaB inhibition. J Biol Chem 273:16446–16452

Rougvie AE, Lis JT (1988) The RNA polymerase II molecule at the 5' end of the uninduced hsp70 gene of D. melanogaster is transcriptionally engaged. Cell 54:795–804

Sandqvist A, Bjork JK, Akerfelt M, Chitikova Z, Grichine A, Vourc'h C, Jolly C, Salminen TA, Nymalm Y, Sistonen L (2009) Heterotrimerization of heat-shock factors 1 and 2 provides a transcriptional switch in response to distinct stimuli. Mol Biol Cell 20:1340–1347

Sarge KD, Murphy SP, Morimoto RI (1993) Activation of heat shock gene transcription by heat shock factor 1 involves oligomerization, acquisition of DNA-binding activity, and nuclear localization and can occur in the absence of stress. Mol Cell Biol 13:1392–1407

Satyal SH, Schmidt E, Kitagawa K, Sondheimer N, Lindquist S, Kramer JM, Morimoto RI (2000) Polyglutamine aggregates alter protein folding homeostasis in Caenorhabditis elegans. Proc Natl Acad Sci U S A 97:5750–5755

Shamovsky I, Gershon D (2004) Novel regulatory factors of HSF-1 activation: facts and perspectives regarding their involvement in the age-associated attenuation of the heat shock response. Mech Ageing Dev 125:767–775

Sharp S, Workman P (2006) Inhibitors of the HSP90 molecular chaperone: current status. Adv Cancer Res 95:323–348

Shi Y, Manley JL (2007) A complex signaling pathway regulates SRp38 phosphorylation and pre-mRNA splicing in response to heat shock. Mol Cell 28:79–90

Shi Y, Mosser DD, Morimoto RI (1998) Molecular chaperones as HSF1-specific transcriptional repressors. Genes Dev 12:654–666

Shin C, Feng Y, Manley JL (2004) Dephosphorylated SRp38 acts as a splicing repressor in response to heat shock. Nature 427:553–558

Sorger PK, Pelham HR (1988) Yeast heat shock factor is an essential DNA-binding protein that exhibits temperature-dependent phosphorylation. Cell 54:855–864

Sorger PK, Lewis MJ, Pelham HR (1987) Heat shock factor is regulated differently in yeast and HeLa cells. Nature 329:81–84

Sprang GK, Brown IR (1987) Selective induction of a heat shock gene in fibre tracts and cerebellar neurons of the rabbit brain detected by in situ hybridization. Brain Res 427:89–93

Stadtman ER (1992) Protein oxidation and aging. Science 257:1220–1224

Takeyama K, Ito S, Yamamoto A, Tanimoto H, Furutani T, Kanuka H, Miura M, Tabata T, Kato S (2002) Androgen-dependent neurodegeneration by polyglutamine-expanded human androgen receptor in Drosophila. Neuron 35:855–864

Taylor IC, Workman JL, Schuetz TJ, Kingston RE (1991) Facilitated binding of GAL4 and heat shock factor to nucleosomal templates: differential function of DNA-binding domains. Genes Dev 5:1285–1298

Teixeira-Castro A, Ailion M, Jalles A, Brignull HR, Vilaca JL, Dias N, Rodrigues P, Oliveira JF, Neves-Carvalho A, Morimoto RI, Maciel P (2011) Neuron-specific proteotoxicity of mutant ataxin-3 in C. elegans: rescue by the DAF-16 and HSF-1 pathways. Hum Mol Genet 20:2996–3009

Trinklein ND, Murray JI, Hartman SJ, Botstein D, Myers RM (2004) The role of heat shock transcription factor 1 in the genome-wide regulation of the mammalian heat shock response. Mol Biol Cell 15:1254–1261

Trott A, West JD, Klaic L, Westerheide SD, Silverman RB, Morimoto RI, Morano KA (2008) Activation of heat shock and antioxidant responses by the natural product Celastrol: transcriptional signatures of a thiol-targeted molecule. Mol Biol Cell 19:1104–1112

Voisin PJ, Pardue S, Macouillard F, Yehia G, Labouesse J, Morrison-Bogorad M (1996) Differential expression of heat shock 70 proteins in primary cultures from rat cerebellum. Brain Res 739:215–234

Wang Z, Lindquist S (1998) Developmentally regulated nuclear transport of transcription factors in Drosophila embryos enable the heat shock response. Development 125:4841–4850

Warrick JM, Paulson HL, Gray-Board GL, Bui QT, Fischbeck KH, Pittman RN, Bonini NM (1998) Expanded polyglutamine protein forms nuclear inclusions and causes neural degeneration in Drosophila. Cell 93:939–949

Westerheide SD, Morimoto RI (2005) Heat shock response modulators as therapeutic tools for diseases of protein conformation. J Biol Chem 280:33097–33100

Westerheide SD, Bosman JD, Mbadugha BN, Kawahara TL, Matsumoto G, Kim S, Gu W, Devlin JP, Silverman RB, Morimoto RI (2004) Celastrols as inducers of the heat shock response and cytoprotection. J Biol Chem 279:56053–56060

Westerheide SD, Anckar J, Stevens SM Jr, Sistonen L, Morimoto RI (2009) Stress-inducible regulation of heat shock factor 1 by the deacetylase SIRT1. Science 323:1063–1066

Whitesell L, Lindquist SL (2005) HSP90 and the chaperoning of cancer. Nat Rev Cancer 5:761–772

Whitesell L, Bagatell R, Falsey R (2003) The stress response: implications for the clinical development of hsp90 inhibitors. Curr Cancer Drug Targets 3:349–358

Williams GT, McClanahan TK, Morimoto RI (1989) E1a transactivation of the human HSP70 promoter is mediated through the basal transcriptional complex. Mol Cell Biol 9:2574–2587

Wu BJ, Williams GT, Morimoto RI (1987) Detection of three protein binding sites in the serum-regulated promoter of the human gene encoding the 70-kDa heat shock protein. Proc Natl Acad Sci U S A 84:2203–2207

Wu C (1995) Heat shock transcription factors: structure and regulation. Annu Rev Cell Dev Biol 11:441–469

Xia W, Guo Y, Vilaboa N, Zuo J, Voellmy R (1998) Transcriptional activation of heat shock factor HSF1 probed by phosphopeptide analysis of factor 32P-labeled in vivo. J Biol Chem 273:8749–8755

Xiao X, Zuo X, Davis AA, McMillan DR, Curry BB, Richardson JA, Benjamin IJ (1999) HSF1 is required for extra-embryonic development, postnatal growth and protection during inflammatory responses in mice. EMBO J 18:5943–5952

Yamamoto A, Mizukami Y, Sakurai H (2005) Identification of a novel class of target genes and a novel type of binding sequence of heat shock transcription factor in Saccharomyces cerevisiae. J Biol Chem 280:11911–11919

Zhang Y, Huang L, Zhang J, Moskophidis D, Mivechi NF (2002) Targeted disruption of hsf1 leads to lack of thermotolerance and defines tissue-specific regulation for stress-inducible Hsp molecular chaperones. J Cell Biochem 86:376–393

Zhou YN, Kusukawa N, Erickson JW, Gross CA, Yura T (1988) Isolation and characterization of Escherichia coli mutants that lack the heat shock sigma factor sigma 32. J Bacteriol 170:3640–3649

Zoghbi HY, Orr HT (2000) Glutamine repeats and neurodegeneration. Annu Rev Neurosci 23:217–247

Zou J, Guo Y, Guettouche T, Smith DF, Voellmy R (1998) Repression of heat shock transcription factor HSF1 activation by HSP90 (HSP90 complex) that forms a stress-sensitive complex with HSF1. Cell 94:471–480

The endoplasmic reticulum unfolded protein response and neurodegeneration

David Ron

Abstract Proteostasis in the lumen of the endoplasmic reticulum is defended by signalling pathways that match the load of unfolded proteins to the complement of chaperones in the organelle. This balancing act is attained via a transcriptional program that activates genes encoding chaperones and other proteins that function in the secretory pathway and a translational program that transiently attenuates protein synthesis by regulating translation initiation. The key features of this unfolded protein response (UPR) will be reviewed, with an emphasis on those strands that are known to intersect with neurophysiology and neuropathology. A particular focus for review will be the evidence that activity of the translational arm of the UPR impacts on diverse phenomena ranging from memory consolidation to myelination and that the transcriptional arm of the UPR is in equilibrium with other pathways that defend proteostasis. Specific consideration will be given to points for intervention in the working of the UPR and how these might be harnessed for treatment of neurological disease.

Overview of the unfolded protein response

Protein secretion is key to cells' interaction with their environment and is at the core of multicellularity. This is conspicuously the case for the immune, endocrine and nervous systems of higher eukaryotes, where intercellular communication depends on the secretion and presentation on the cell surface of multiple protein ligands and receptors. Secreted and transmembrane proteins undergo early steps of their biogenesis in the lumen of the endoplasmic reticulum (ER), where a host of chaperones and protein-modifying enzymes assist the unfolded nascent chain to attain its

D. Ron (✉)
University of Cambridge Metabolic Research Laboratories and NIHR Cambridge Biomedical Research Centre, MRL, level 4, Addenbrooke's Hospital, Box 289, Hills Road, Cambridge CB2 0QQ, UK
e-mail: dr360@medschl.cam.ac.uk

R.I. Morimoto and Y. Christen (eds.), *Protein Quality Control in Neurodegenerative Diseases*, Research and Perspectives in Alzheimer's Disease,
DOI 10.1007/978-3-642-27928-7_2,

proper three-dimensional and oligomeric structure. The ER maintains a quality control function, retaining unfolded or misfolded proteins and selectively allowing only properly folded and assembled proteins to egress from the organelle and progress down the secretory pathway (Ellgaard and Helenius 2003). Proteins persistently engaged by this quality control machinery are eventually translocated back into the cytosol and degraded in a process known as ER associated protein degradation (ERAD; Tamura et al. 2010).

The mass of unfolded proteins translocated into the ER is modulated by diverse physiological exigencies. Furthermore, physiological variables impact on the functional state of the protein folding machinery in the ER lumen. To cope with these changing circumstances, eukaryotes have evolved dedicated signal transduction pathways that couple the capacity of the ER protein folding machinery and the load of unfolded proteins. These can be triggered experimentally by manipulations that promote protein misfolding in the ER and consequently have been named the unfolded protein response(s), or UPR (Kozutsumi et al. 1988).

Whilst there is a debate remaining in regard to the molecular mechanisms by which the UPR signal transducers are activated, it is clear that, directly or indirectly, the balance between unfolded proteins and chaperones is sensed (Kimata and Kohno 2011). Thus, an imbalance favoring unfolded proteins defines ER stress. It also follows that cells experience periodic low levels of ER stress under quite normal circumstances and that the study of its consequences is as relevant to physiology as it is to pathophysiology.

Animal cells respond to ER stress with two distinct adaptations: a rapid and transient reduction in rates of new protein synthesis and a more gradual alteration of gene expression programs (Ron and Walter 2007; Fig. 1). The former is an immediate rectifying response that indiscriminately attenuates the flux of new proteins entering the ER, whereas the latter globally up-regulates genes encoding proteins that function at all levels of the secretory pathway. The gene expression arm of UPR, which is conserved in all eukaryotes, enhances the ability of cells to cope with their secretory mission. This gene expression program, which includes many ER chaperones and enzymes, represents a signature of the UPR and serves to distinguish it from other stress pathways (Travers et al. 2000; Murray et al. 2004).

The translational arm of the UPR

PERK

ER stress is linked to attenuated rates of translation initiation by an ER localized transmembrane protein, PERK (Harding et al. 1999). PERK's lumenal stress-sensing domain monitors the balance of unfolded proteins and chaperones and

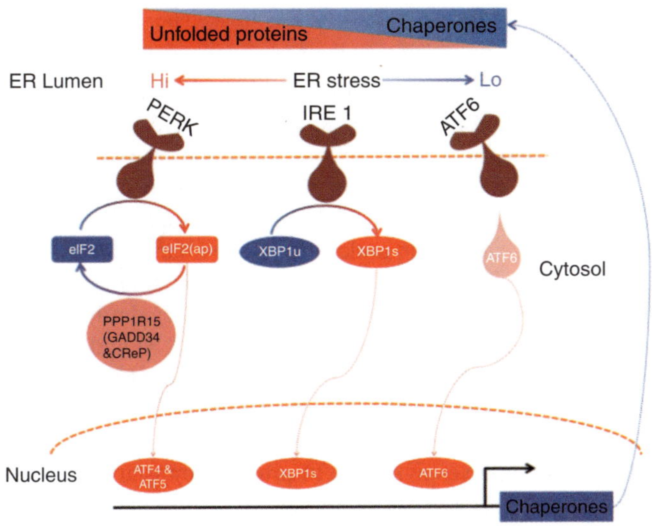

Fig. 1 The endoplasmic reticulum unfolded protein response (UPR). The balance between chaperones and unfolded proteins in the ER lumen establishes the levels of ER stress, which is recognized by three known transmembrane stress receptors. The translation strand of the UPR is mediated single handedly by the transmembrane eIF2α kinase PERK (*left*). Phosphorylation of eIF2α attenuates new protein synthesis, lowering the supply of unfolded proteins to the ER while, at the same time promoting the translation of the transcription factors ATF4 and ATF5. IRE1 signalling (*center*) promotes the unconventional splicing of the XBP1 mRNA, converting it from the inactive unspliced (XBP1u) form to the XBP1s form, encoding a transcriptional activator. ATF6 (*right*) is cleaved in ER stressed cells to generate an active transcription factor that, together with ATF4, ATF5 and XBP1, activates UPR target genes to enhance the capacity of the ER to cope with unfolded proteins

promotes oligomerization of PERK in the plane of the membrane in response to ER stress. PERK's cytosolic effector domain is a protein kinase that undergoes activating trans-autophosphorylation by this clustering. Activated PERK recruits and phosphorylates translation initiation factor 2 (eIF2) on serine 51 of its alpha subunit (Harding et al. 1999; Marciniak et al. 2006). In its GTP-bound form, eIF2 plays an essential role in translation initiation, and the phosphorylation of its alpha subunit inhibits the exchange of GDP for GTP on eIF2, inhibiting rates of protein synthesis (Hinnebusch 2000). Thus PERK single-handedly couples ER stress to attenuated rates of protein synthesis (Harding et al. 2000a).

In the absence of PERK, the flux of proteins into the ER is no longer responsive to the levels of ER stress, exposing cells to enhanced risk of protein misfolding in the ER and promoting higher levels of activity in the parallel strands of the UPR (Harding et al. 2000a). The loss of ER stress-regulated eIF2α phosphorylation is an especially great problem for professional secretory cells, leading to impairment of the endocrine and exocrine pancreas, salivary glands, liver and bone (Harding et al. 2001; Scheuner et al. 2001; Zhang et al. 2002).

The integrated stress response

In addition to negatively regulating rates of protein synthesis, eIF2α phosphorylation affects gene transcription (Harding et al. 2000b). This transcription is attained by the seemingly paradoxical translational upregulation of transcription factors such as ATF4 and ATF5, under conditions of high levels of phosphorylated eIF2α, abbreviated eIF2(αP) (Lu et al. 2004a; Vattem and Wek 2004; Zhou et al. 2008). A detailed discussion of the underlying molecular mechanism is beyond the scope of this review. It is, however, worth noting that guanine nucleotide exchange is the focal point of eIF2(αP) action, both in terms of regulating global translation and in terms of transcriptional activation (via ATF4, ATF5 and similarly regulated factors). eIF2(αP) inhibits the guanine nucleotide exchange activity of the GEF, eIF2B (Siekierka et al. 1981; Clemens et al. 1982) and strengthens the anti-exchange (GDI) activity of eIF5 (Jennings and Pavitt 2010), leading to lower levels of GTP-bound eIF2 complexes available to promote translation initiation.

Whilst PERK is the only kinase to couple eIF2α phosphorylation to ER stress, cells have three other kinases that affect eIf2α phosphorylation in response to other stresses: GCN2 to amino acid starvation, PKR to viral infection and HRI to heme depletion. Thus, eIF2(αP) integrates signaling in several stress responses and the subordinate gene expression program is referred to as the integrated stress response (ISR). The ISR constitutes one strand of the UPR, but it makes an important contribution to the overall gene expression program activated by ER stress, with genes involved in resistance to oxidative stress as particularly important ISR targets (Harding et al. 2003). It has been reported that PERK couples ER stress to phosphorylation and activation of NRF2, a transcription factor with a prominent role in resistance to oxidative stress (Cullinan et al. 2003); however, PERK-dependent gene expression is completely abolished in fibroblasts by mutating serine 51 of eIF2α to alanine (Lu et al. 2004b). Therefore, it would appear that PERK employs a rather private pathway, utilizing a single substrate to promote ER stress signaling.

eIF2(αP) dephosphorylation

Phosphorylation of eIF2α is reversed by PP1 phosphatase complexes that are directed to eIF2 by one of two regulatory subunits: the constitutively expressed PPP1R15B (also known as CReP; Jousse et al. 2003) and the ISR-inducible PPP1R15A (also known as GADD34; Novoa et al. 2001; Brush et al. 2003). The latter functions in a homeostatic negative feed-back loop that restores translation in stressed cells. Consequently, PPP1R15A/GADD34-deficient cells have higher and more sustained levels of eIF2(αP) with delayed translational recovery (Novoa et al. 2003). Deletion of the constitutive PPP1R15B/CReP leads to perinatal lethality in mice (Harding et al. 2009), but mice lacking PPP1R15A/GADD34 appear healthy.

Interestingly, the PPP1R15A/GADD34 mutant mice and cells derived from them are partially resistant to the lethal effects of toxins that promote ER stress (Marciniak et al. 2004), a phenotype shared by mice lacking the transcription factor CHOP (Zinszner et al. 1998). CHOP is transcriptionally induced by the ISR and has PPP1R15A/GADD34 as one of its target genes (Marciniak et al. 2004).

These observations suggest that the pathway for translational recovery mediated by eIF2(αP) dephosphorylation is subject to failure of homeostasis, in that attenuating this negative feedback loop has beneficial effects in some circumstances. This feature of the ISR also explains the recently described survival-promoting effects of guanabenz (Wytensin™) when applied (at relatively high concentrations) to ER-stressed cells. Guanabenz was found to selectively disrupt the interaction of PPP1R15A/GADD34 with the PP1 catalytic subunit, enforcing in effect a GADD34-deficient state (Tsaytler et al. 2011). The basis for the selectivity for PPP1R15A/GADD34 over PPP1R15B/CReP is not understood, but it is a critical feature of the compound: whereas wildtype cells tolerate guanabenz well, PPP1R15B/CReP knock-out cells exposed to guanabenz succumb to mounting levels of eIF2(αP) (Tsaytler et al. 2011), a phenotype predicted by the lethality of combined deletion of the two eIF2(αP)-directed phosphatase regulatory subunits (Harding et al. 2009).

The genetic and pharmacological experiments described above show that extremes of eIF2(αP) levels are poorly tolerated. Cells lacking PERK are presumably at excess risk of protein misfolding and ER-based proteotoxicity and cells lacking eIF2(αP)-directed phosphatases are unable to sustain levels of protein synthesis. However, between these two extremes there is quite some latitude for manipulation, a point to return to later when considering the role of the ISR in neuropathophysiology.

The transcriptional arm of the UPR

IRE1

The oldest and most conserved arm of the UPR is mediated by IRE1, an unusual ER localized transmembrane protein (Cox et al. 1993; Mori et al. 1993). IRE1 shares with its descendant, PERK, a functionally-interchangeable stress-sensing lumenal domain. IRE1's cytosolic effector domain is also a protein kinase; however, its only known substrates are other IRE1 molecules. In response to ER stress, IRE1 clusters in the plane of the membrane and undergoes activating *trans*-autophosphorylation (Shamu and Walter 1996). Phosphorylation unmasks a second catalytic function of IRE1: to cleave a pre-existing mRNA, HAC1 in yeast (Cox and Walter 1996) and XBP1 in animals, at precisely two locations (Yoshida et al. 2001; Calfon et al. 2002), liberating a small fragment (26 bases in mammals). The two ends of the XBP1 mRNA are joined together to complete the unconventional splicing event, which results in a frame shift. The spliced XBP1 mRNA encodes a potent transcription factor that activates UPR target genes.

In addition to this highly conserved and unusual signal transduction pathway, IRE1 has been reported to engage in two other activities: phosphorylated mammalian IRE1 can recruit the TRAF2 adaptor, activating Jun N-terminal kinase (Urano et al. 2000; Nishitoh et al. 2002) or caspases (Yoneda et al. 2001). This strand of UPR activity has been proposed to link ER stress to cell death and dysfunction, but the physiological significance of these links has not been critically explored. IRE1 also possesses promiscuous RNase activity directed to ER membrane bound mRNA. Referred to as regulated IRE1-dependent degradation (or RIDD), it is believed to work alongside PERK in attenuating the load of unfolded proteins that enter the ER by degrading their encoding mRNAs (Hollien and Weissman 2006; Hollien et al. 2009). At present, there are no good genetic or pharmacological tools to separate XBP1 splicing from RIDD; therefore, it has not been possible to critically examine RIDD's contribution to protein folding homeostasis. However, observations made in an IRE1 over-expression model suggest that RIDD may contribute to the death of ER-stressed insulinoma cells (Han et al. 2009).

IRE1's kinase activity and RNase activity are mediated by different portions of the cytosolic effector domain and are separable (Lee et al. 2008). In vitro work with the yeast enzyme shows the importance of ligand engagement at the nucleotide binding pocket, not only to the kinase activity (which is expected) but also to the RNAse activity (Papa et al. 2003). The two are likely coupled by a ligand binding-dependent dimerization event (Lee et al. 2008), which may extend to a higher order, more active oligomer (Korennykh et al. 2009). Additional evidence for allosteric regulation of IRE1 is provided by the ability of the flavonol quercetin to activate yeast IRE1 in vitro by engaging a ligand-binding pocket at the dimer interface, which stabilizes the active dimeric form of the enzyme (Wiseman et al. 2010).

It remains to be seen to what extent these lessons from the yeast enzyme are applicable to mammalian IRE1. But the prospects of being able to tune and bias IRE1's signaling by ligands that engage its various allosteric and active sites are very enticing.

ATF6

The third known strand of the UPR is mediated by ATF6. In unstressed cells, ATF6 is found as a latent transcription factor, tethered to the ER membrane. However, under conditions of ER stress, regulated intra-membrane proteolysis liberates ATF6 from the ER membrane and the soluble transcription factor migrates to the nucleus to join ATF4 and the other mediators of the ISR and XBP1 in activating UPR target genes (Yoshida et al. 1998; Haze et al. 1999; Ye et al. 2000).

The proximate antecedent of ATF6 cleavage and activation is its migration from the ER to the Golgi. Once at the Golgi, ATF6 is cleaved by the same proteases that activate the sterol-response element binding proteins (SREBPs; Ye et al. 2000).

ATF6 is co-activated with the other strands of the UPR, and BiP binding has been proposed to play a role in repressing its trafficking to the Golgi in unstressed cells (Shen et al. 2002). Mammals have two ATF6 genes, deletion of which significantly compromises resistance to ER stress (Yamamoto et al. 2007). In addition, mammals possess several other proteins with similar architecture and (likely) similar activation mechanisms (e.g., CREB-H and OASIS). Their role in the ER stress response is currently under study in several labs.

UPR signaling in neurophysiology

Gross brain development is tolerant of deletion of key branches of the UPR (unlike the sensitivity of secretory organs to such genetic manipulation). PERK knockout mice and humans with mutations in the PERK-encoding *EIF2AK3* gene (leading to the Walcott-Rallison syndrome) have relatively normal CNS function (Harding et al. 2001), whereas conditional deletion of XBP1 in brain tissue is also compatible with normal development of the organ (Hetz et al. 2009). However, before considering the role of the UPR in neuropathophysiology it is worth noting two interesting contributions that regulated eIF2α phosphorylation makes to neurophysiology.

eIF2(αP), feeding behavior and memory formation

In yeast, the single eIF2α kinase GCN2 couples amino acid availability to an eIF2 (αP)-dependent gene expression program, akin to the ISR. The gene products of this program adapt the yeast to an environment limited in amino acids (Dever et al. 1992). A conceptually similar response is induced by GCN2 in the piriform cortex of mice fed a diet deficient in a single amino acid (Hao et al. 2005; Maurin et al. 2005). The basis of GCN2 activation by amino acid deprivation is fairly well understood; however, the molecular events triggered by GCN2-mediated eIF2 (αP) in the responding neurones are not understood and likely involve both transcriptional and post-transcriptional events.

The regulated phosphorylation of eIF2α also plays an important role in memory formation. In brain slices of mice lacking GCN2, long-term potentiation by a single train of electrical stimulation is enhanced, compared with the wildtype (Costa-Mattioli et al. 2005). This likely hinges on the interplay between the eIF2(αP)-dependent expression of ATF4 and the related transcription factor CREB, but neither the upstream mediators of GCN2 activation, in this context, nor the downstream events that connect eIF2(αP) to synaptic plasticity are known.

The uncertainty as regards to mechanism notwithstanding, the experimental indictment of eIF2(αP) in fundamental aspects of neurophysiology has an important implication as we consider the role of the UPR in neuropathology because it suggests that the phenotypic consequences of UPR signaling, of which eIF2α phosphorylation is an important component, may be established not only through effects on cellular health and viability but also through the impact of UPR signaling on functional circuitry.

UPR signaling in neuropathophysiology

Control of translation initiation and the syndrome of childhood ataxia and cerebral hypomyelination (CACH)

Rare, loss-of-function mutations affecting the guanine nucleotide exchange factor eIF2B are strongly associated with a syndrome of episodic progressive childhood ataxia with CNS hypomyelination (Fogli and Boespflug-Tanguy 2006; Schiffmann and Elroy-Stein 2006). eIF2B activity is a target for repression by phosphorylated eIF2α, which mediates the effects of eIF2(αP) on both global translation and on gene-specific induction in the ISR. As expected, fibroblasts of individuals with CACH have heightened activity in their ISR when exposed to conditions that impose mild levels of ER stress (Kantor et al. 2005).

The notion that a misregulated ISR contributes to the pathophysiology of CACH is further supported by the clinical history. Patients are reported to experience bouts of demyelination in the context of acute febrile illness or head trauma, events that are likely to enhance ER stress signaling and promote transient increases in levels of eIF2(αP). The notion that CACH is a disease caused by a de-regulated ISR remains an unproven hypothesis but, if correct, may have interesting implications for therapy for diseases of aging and protein misfolding.

As noted above, loss-of-function mutations in PPP1R15A/GADD34 are associated with resistance to cell death and dysfunction caused by ER stress. This is attributed to a sustained induction of eIF2(αP) levels in the mutant and attenuated unfolded protein load confronting the ER. A recent study suggests that the potential benefits of sustained ISR activity may extend to protein misfolding in other compartments, too. The anti-hypertensive drug Guanabenz (Wytensin™) was found to have anti-prion activity (Tribouillard-Tanvier et al. 2008) and also to protect cells against ER stress. The latter activity is well explained by the selective dissociation of PPP1R15A/GADD34 from the catalytic subunit of the phosphatase by Guanabenz (Tsaytler et al. 2011). This observation and previous reports that Salubrinol, a compound that inhibits eIF2(αP) dephosphorylation by yet-to-be determined mechanism(s), protects against ER stress (Boyce et al. 2005) nominate GADD34 as a target for inhibition to treat/prevent diseases of protein misfolding. It remains to be determined whether the protection afforded by Guanabenz against prions is also mediated by inhibition of GADD34.

Together, these disparate strands suggest that protein folding is subject to a failure of homeostasis that can be rectified, in some circumstances, by inhibition of GADD34 and an accentuated and sustained ISR. According to one very simple

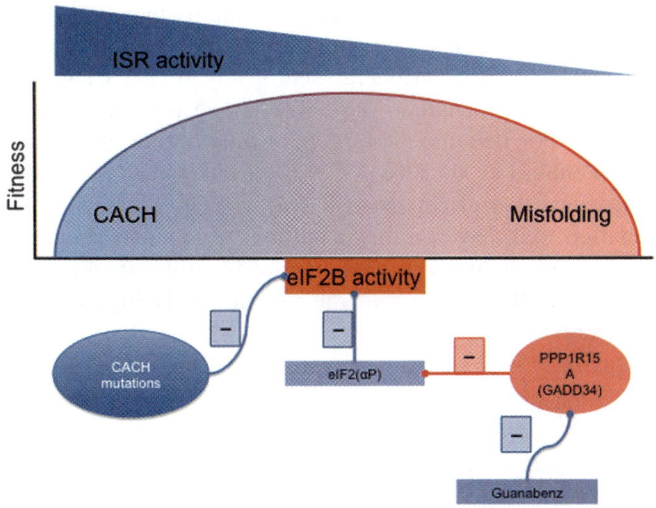

Fig. 2 Extremes of ISR signalling are detrimental. Shown is a schema of the hypothesized relationship between fitness (in terms of health of CNS myelin) and ISR activity, which is regulated by eIF2B activity. The latter is attenuated both by mutations associated with the syndrome of childhood ataxia and cerebral hypomyelination (CACH) and by elevated eIF2α phosphorylation. The PPP1R15A/GADD34 phosphatase regulatory subunit promotes eIF2(αP) dephosphorylation and attenuates ISR signalling. This action can be reversed by Guanabenz, which has recently been demonstrated to protect cells from ER stress (Tsaytler et al. 2011)

model, CACH reports on the costs of an inappropriately robust and sustained ISR, paid for by a catastrophic collapse of myelin homeostasis (Fig. 2). If these ideas are correct, CACH defines one of the limitations of the benefits of a therapeutic strategy aimed at strengthening the ISR.

Inflammatory CNS demyelination and the ISR

The ISR also appears to play an important role in the defense against inflammatory demyelination in the CNS. In this context, it is an insufficiently robust ISR (imposed by haploid insufficiency of PERK) that exposes mice to enhanced demyelination in a model of experimental allergic encephalitis (EAE; Lin et al. 2007) or cytokine-induced CNS inflammation (Lin et al. 2006), and GADD34 deletion protects against oligodendrocyte loss in these models (Lin et al. 2008).

It is unclear if CACH and the EAE/cytokine-induced CNS inflammation are exposing different facets of a special role for the ISR in myelin homeostasis. In the case of the CACH syndrome, it is not even clear in which cell type(s) the drama of attenuated eIF2B activity is played out, while the role of the ISR in inflammatory demyelination is likely played out at the level of the myelin producing cell (the oligodendrocyte, in case of the CNS). The latter is supported by evidence from the peripheral nervous system, where mutations that lead to misfolding of abundantly expressed myelin constituents affect the viability and health of the Schwann cell and lead to a peripheral neuropathy of the Charcot-Marie-Tooth type. In a mouse model of Charcot-Marie-Tooth type 1B, caused by a misfolding-prone mutation in P0, deletion of CHOP restores motor function and reduces demyelination (Pennuto et al. 2008). It will be interesting to learn what role, if any, other mutations that affect the ISR (e.g., PERK haploid insufficiency and GADD34 mutation) will have on this condition. A coherent phenotype, with ISR loss-of-function exacerbating and ISR gain-of-function ameliorating demyelination, would support the notion that the myelin-producing cells are especially sensitive to ER stress.

Parkinson's disease, defective vesicular transport and chaperone inactivation

Evidence of enhanced UPR activity in the affected brain regions of Parkinson's disease (PD) patients and of patients suffering from the related multi-systems atrophy syndrome has been noted (Hoozemans et al. 2007; Makioka et al. 2010), and experimental manipulations that model aspects of PD in cultured cells and mouse tissues have been noted to promote UPR signaling (Ryu et al. 2002; Holtz and O'Malley 2003; Silva et al. 2005). This finding engenders two important questions: how are ER stress signaling pathways engaged in PD, and what roles do the underlying ER stress or the response to it play in the pathophysiology of PD?

Studies in yeast have shown that synuclein over-expression interferes with ER to Golgi transport of vesicles (Cooper et al. 2006), an effect that appears to be direct, as it can be reproduced in an in vitro transport assay (Gitler et al. 2008). The defect can be corrected by over-expression of Rab1, a small GTPase that regulates aspects of vesicular transport. Remarkably, Rab1 over-expression also attenuates the dysfunction of dopaminergic neurons over-expressing synuclein (Cooper et al. 2006), supporting a role for defective vesicular transport in the toxicity of synuclein. But is this related to ER stress?

ER to Golgi transport is key to secretion and interfering with it may compromise diverse cellular functions, besides imposing a measure of ER stress (presumably due to back-up of proteins in the ER). Thus, it is unclear if the ER misfolded protein stress that likely accompanies a mutant synuclein expression-mediated block to vesicular transport (Smith et al. 2005) contributes to the pathophysiology of PD. The point would be mooted by interventions that rectify the underlying defect; therefore, coming up with such interventions seems a priority. If this proves challenging, then it may be worth pausing to consider the role of ER stress and ER stress signaling in the pathophysiology of PD.

Altered vesicular transport may not be the only mechanism linking PD with ER stress; inactivating S-nitrosation of the ER-localized protein disulfide isomerase has been reported in brain samples from individuals with sporadic PD (Uehara et al. 2006). This post-translational modification is predicted to affect disulfide bond formation and re-shuffling and thus has the potential to compromise protein folding in the ER and promote ER stress signaling.

Genetic studies point to a role for ER stress in the demise of dopaminergic neurons in toxigenic models of PD in mice. CHOP deficiency protects the neonatal striatum from 6-hydroxydopamine (Silva et al. 2005), whereas targeted expression of spliced XBP1 protected dopaminergic neurons against 1-methyl-4-phenylpyridinium (MPP+) in vitro (Sado et al. 2009) and ATF6α deficiency sensitized the striatum to similar insults in vivo (Egawa et al. 2010). These experiments strongly support a role for ER stress in the death and dysfunction of dopaminergic neurons exposed to toxigenic models of PD; however, the relevance of ER stress and the response to it in the pathogenesis of human PD remain unclear.

The UPR and cytosolic proteotoxicity

One of the themes to emerge from the study of PD is that pathological processes that occur in the cytosol, like the misfolding of α-synuclein, may have consequences on the ER and lead to (potentially) significant ER misfolded protein stress. This theme is echoed in cellular models of other neurodegenerative diseases. Elevated levels of ER stress markers have been noted in yeast (Bence et al. 2001) and mammalian cells (Nishitoh et al. 2002) over-expressing misfolding-prone poly-glutamine expanded proteins, like Huntingtin. This finding correlates with dysfunction of the ubiquitin proteasome system and suggests a model whereby attenuated degradation of misfolded/unfolded ER proteins (an ER-associated degradation, or ERAD, defect) triggers secondary ER stress in what is primarily a state of cytoplasmic proteotoxicity (Duennwald and Lindquist 2008).

The notion that normal function of the cytosolic ubiquitin-proteasome system is limiting to ERAD and ER protein folding homeostasis is further supported by high levels of ER stress noted in secretory cells exposed to proteasome inhibitors — such as Bortezomib, used in cancer therapy (Nawrocki et al. 2005; Obeng et al. 2006). A role for ER stress in the death and dysfunction of cells expressing misfolding-prone poly-glutamine expanded proteins is suggested by the observation that, in a cellular model of Huntington's Disease, disrupting the link between ER stress and Jun N-terminal kinase by inactivation of the upstream ASK1 kinase ameliorates proteotoxicity (Nishitoh et al. 2002). However, the extent to which these observations would carry over to tissue models, let alone the human disease, is unclear. Furthermore, it is unclear if the benefits of ASK1 inactivation accrue solely via disruption of this ER nexus.

An intriguing development in the study of the links between ER stress and cytoplasmic proteotoxicity concerns the case of a mouse model of amyotrophic lateral sclerosis (ALS)/ motor neuron disease, based on targeted expression of an ALS-associated mutant $SOD1^{G86R}$. SOD1 is not a secreted protein; its misfolding-prone mutant versions accumulate intracellularly and, like the aforementioned cytosolic proteotoxicity diseases, experimental models of both ALS and the sporadic disease in humans are associated with activation of UPR markers in the motor neurons (Atkin et al. 2008). Therefore, the conceptual frame of a transmembrane equilibrium of proteostasis — whereby protein misfolding in the cytosol begets ER stress via attenuated ERAD — predicted that further compromise of the UPR would accelerate disease development caused by mutant SOD1 and enhance its severity. The opposite was observed. Deletion of XBP1 in the nervous system delayed degeneration of motor neurons in an $SOD1^{G86R}$ transgenic model (Hetz et al. 2009). This surprising observation correlated with enhanced clearance of the mutant SOD1 by autophagy and suggested a circuitous model whereby defective induction ERAD genes in the XBP1 mutant leads to a compensatory upregulation of autophagy. Mutant SOD1 is apparently caught up in this backup proteostatic mechanism as a bystander and its clearance is enhanced (Hetz et al. 2009).

The pathways linking ER stress to autophagy are incompletely worked up. In yeast, HOG1 links ER stress to autophagy (Bicknell et al. 2010) and, in mammalian cells, induction of autophagy by proteasome inhibitors is IRE1 — but not XBP1-dependent (Ding et al. 2007). Viewed narrowly, these observations suggest that inhibitors of IRE1's sequence-specific RNase activity may delay motor neuron loss in patients with SOD1 mutations. But the broader implication is that the extensive homeostatic feedback in the UPR makes it difficult to predict the consequences of manipulations that enhance or enfeeble specific strands of the response. At the same time, it suggests that tools to selectively (and transiently) inhibit UPR signaling may be applied to mobilize homeostatic responses to intervene in the pathophysiology of neurodegeneration. Because UPR signaling is conducted via rather private pathways through the activity of enzymes like IRE1 and PERK with few substrates, it presents several attractive drug targets with potential for considerable selectivity. It will be interesting to learn if efforts currently underway to exploit these nascent concepts will prove relevant to the treatment of neurodegenerative diseases.

References

Atkin JD, Farg MA, Walker AK, McLean C, Tomas D, Horne MK (2008) Endoplasmic reticulum stress and induction of the unfolded protein response in human sporadic amyotrophic lateral sclerosis. Neurobiol Dis 30:400–407

Bence NF, Sampat RM, Kopito RR (2001) Impairment of the ubiquitin-proteasome system by protein aggregation. Science 292:1552–1555

Bicknell AA, Tourtellotte J, Niwa M (2010) Late phase of the endoplasmic reticulum stress response pathway is regulated by hog1 map kinase. J Biol Chem 285:17545–17555

Boyce M, Bryant KF, Jousse C, Long K, Harding HP, Scheuner D, Kaufman RJ, Ma D, Coen D, Ron D, Yuan J (2005) A selective inhibitor of eif2a dephosphorylation protects cells from er stress. Science 307:935–939

Brush MH, Weiser DC, Shenolikar S (2003) Growth arrest and DNA damage-inducible protein gadd34 targets protein phosphatase 1alpha to the endoplasmic reticulum and promotes dephosphorylation of the alpha subunit of eukaryotic translation initiation factor 2. Mol Cell Biol 23:1292–1303

Calfon M, Zeng H, Urano F, Till JH, Hubbard SR, Harding HP, Clark SG, Ron D (2002) Ire1 couples endoplasmic reticulum load to secretory capacity by processing the *xbp-1* mrna. Nature 415:92–96

Clemens MJ, Pain VM, Wong ST, Henshaw EC (1982) Phosphorylation inhibits guanine nucleotide exchange on eukaryotic initiation factor 2. Nature 296:93–95

Cooper AA, Gitler AD, Cashikar A, Haynes CM, Hill KJ, Bhullar B, Liu K, Xu K, Strathearn KE, Liu F, Cao S, Caldwell KA, Caldwell GA, Marsischky G, Kolodner RD, Labaer J, Rochet JC, Bonini NM, Lindquist S (2006) Alpha-synuclein blocks er-golgi traffic and rab1 rescues neuron loss in Parkinson's models. Science 313:324–328

Costa-Mattioli M, Gobert D, Harding HP, Herdy B, Azzi M, Bruno M, Ben Mamou C, Marcinkiewicz E, Yoshida M, Imataka H, Cuello AC, Seidah N, Sossin W, Lacaille J-C, Ron D, Nader K, Sonenberg N (2005) Translational control of hippocampal synaptic plasticity and memory by an eif2 kinase, gcn2. Nature 436:1166–1173

Cox JS, Walter P (1996) A novel mechanism for regulating activity of a transcription factor that controls the unfolded protein response. Cell 87:391–404

Cox JS, Shamu CE, Walter P (1993) Transcriptional induction of genes encoding endoplasmic reticulum resident proteins requires a transmembrane protein kinase. Cell 73:1197–1206

Cullinan SB, Zhang D, Hannink M, Arvisais E, Kaufman RJ, Diehl JA (2003) Nrf2 is a direct perk substrate and effector of perk-dependent cell survival. Mol Cell Biol 23:7198–7209

Dever TE, Feng L, Wek RC, Cigan AM, Donahue TF, Hinnebusch AG (1992) Phosphorylation of initiation factor 2 alpha by protein kinase gcn2 mediates gene-specific translational control of gcn4 in yeast. Cell 68:585–596

Ding W-X, Ni H-M, Gao W, Yoshinori T, Stolz DB, Ron D, Yin X-M (2007) Linking of autophagy to ubiquitin-proteasome system is important for the regulation of endoplasmic reticulum stress and cell viability. Am J Pathol 171:513–524

Duennwald ML, Lindquist S (2008) Impaired erad and er stress are early and specific events in polyglutamine toxicity. Genes Dev 22:3308–3319

Egawa N, Yamamoto K, Inoue H, Hikawa R, Nishi K, Mori K, Takahashi R (2010) The endoplasmic reticulum stress sensor, atf6α protects against neurotoxin-induced dopaminergic neuronal death. J Biol Chem 286:7947–7957

Ellgaard L, Helenius A (2003) Quality control in the endoplasmic reticulum. Nat Rev Mol Cell Biol 4:181–191

Fogli A, Boespflug-Tanguy O (2006) The large spectrum of eif2b-related diseases. Biochem Soc Trans 34:22–29

Gitler AD, Bevis BJ, Shorter J, Strathearn KE, Hamamichi S, Su LJ, Caldwell KA, Caldwell GA, Rochet JC, McCaffery JM, Barlowe C, Lindquist S (2008) The Parkinson's disease protein alpha-synuclein disrupts cellular rab homeostasis. Proc Natl Acad Sci USA 105:145–150

Han D, Lerner AG, Vande Walle L, Upton JP, Xu W, Hagen A, Backes BJ, Oakes SA, Papa FR (2009) Ire1alpha kinase activation modes control alternate endoribonuclease outputs to determine divergent cell fates. Cell 138:562–575

Hao S, Sharp JW, Ross-Inta CM, McDaniel BJ, Anthony TG, Wek RC, Cavener DR, McGrath BC, Rudell JB, Koehnle TJ, Gietzen DW (2005) Uncharged trna and sensing of amino acid deficiency in mammalian piriform cortex. Science 307:1776–1778

Harding H, Zhang Y, Ron D (1999) Translation and protein folding are coupled by an endoplasmic reticulum resident kinase. Nature 397:271–274

Harding H, Zhang Y, Bertolotti A, Zeng H, Ron D (2000a) *Perk* is essential for translational regulation and cell survival during the unfolded protein response. Mol Cell 5:897–904

Harding H, Novoa I, Zhang Y, Zeng H, Wek RC, Schapira M, Ron D (2000b) Regulated translation initiation controls stress-induced gene expression in mammalian cells. Mol Cell 6:1099–1108

Harding H, Zeng H, Zhang Y, Jungreis R, Chung P, Plesken H, Sabatini D, Ron D (2001) Diabetes mellitus and exocrine pancreatic dysfunction in perk−/− mice reveals a role for translational control in survival of secretory cells. Mol Cell 7:1153–1163

Harding H, Zhang Y, Zeng H, Novoa I, Lu P, Calfon M, Sadri N, Yun C, Popko B, Paules R, Stojdl D, Bell J, Hettmann T, Leiden J, Ron D (2003) An integrated stress response regulates amino acid metabolism and resistance to oxidative stress. Mol Cell 11:619–633

Harding HP, Zhang Y, Scheuner D, Chen JJ, Kaufman RJ, Ron D (2009) Ppp1r15 gene knockout reveals an essential role for translation initiation factor 2 alpha (eif2alpha) dephosphorylation in mammalian development. Proc Natl Acad Sci USA 106:1832–1837

Haze K, Yoshida H, Yanagi H, Yura T, Mori K (1999) Mammalian transcription factor atf6 is synthesized as a transmembrane protein and activated by proteolysis in response to endoplasmic reticulum stress. Mol Biol Cell 10:3787–3799

Hetz C, Thielen P, Matus S, Nassif M, Court F, Kiffin R, Martinez G, Cuervo AM, Brown RH, Glimcher LH (2009) Xbp-1 deficiency in the nervous system protects against amyotrophic lateral sclerosis by increasing autophagy. Genes Dev 23:2294–2306

Hinnebusch AG (2000) Mechanism and regulation of initiator methionyl-trna binding to ribosomes. In: Sonenberg N, Hershey JWB, Mathews MB (eds) Translational control of gene expression. CSHL Press, Cold Spring Harbor, pp 185–243

Hollien J, Weissman JS (2006) Decay of endoplasmic reticulum-localized mrnas during the unfolded protein response. Science 313:104–107

Hollien J, Lin JH, Li H, Stevens N, Walter P, Weissman JS (2009) Regulated ire1-dependent decay of messenger rnas in mammalian cells. J Cell Biol 186:323–331

Holtz WA, O'Malley KL (2003) Parkinsonian mimetics induce aspects of unfolded protein response in death of dopaminergic neurons. J Biol Chem 278:19367–19377

Hoozemans JJ, van Haastert ES, Eikelenboom P, de Vos RA, Rozemuller JM, Scheper W (2007) Activation of the unfolded protein response in Parkinson's disease. Biochem Biophys Res Commun 354:707–711

Jennings MD, Pavitt GD (2010) Eif5 has gdi activity necessary for translational control by eif2 phosphorylation. Nature 465:378–381

Jousse C, Oyadomari S, Novoa I, Lu PD, Zhang Y, Harding HP, Ron D (2003) Inhibition of a constitutive translation initiation factor 2a phosphatase, crep, promotes survival of stressed cells. J Cell Biol 163:767–775

Kantor L, Harding HP, Ron D, Schiffmann R, Kaneski CR, Kimball SR, Elroy-Stein O (2005) Heightened stress response in primary fibroblasts expressing mutant eif2b genes from cach/vwm leukodystrophy patients. Hum Genet 118:99–106

Kimata Y, Kohno K (2011) Endoplasmic reticulum stress-sensing mechanisms in yeast and mammalian cells. Curr Opin Cell Biol 23:135–142

Korennykh AV, Egea PF, Korostelev AA, Finer-Moore J, Zhang C, Shokat KM, Stroud RM, Walter P (2009) The unfolded protein response signals through high-order assembly of ire1. Nature 457:687–693

Kozutsumi Y, Segal M, Normington K, Gething MJ, Sambrook J (1988) The presence of malfolded proteins in the endoplasmic reticulum signals the induction of glucose-regulated proteins. Nature 332:462–464

Lee KPK, Dey M, Neculai D, Cao C, Dever TE, Sicheri F (2008) Structure of the dual enzyme ire1 reveals the basis for catalysis and regulation in non-conventional rna splicing. Cell 132:89–100

Lin W, Kemper A, Dupree J, Harding H, Ron D, Popko B (2006) Interferon-γ inhibits central nervous system remyelination through a process modulated by er stress. Brain 129:1306–1318

Lin W, Bailey SL, Ho H, Harding HP, Ron D, Miller SD, Popko B (2007) The integrated stress response prevents demyelination by protecting oligodendrocytes against immune-mediated damage. J Clin Invest 117:448–458

Lin W, Kunkler PE, Harding HP, Ron D, Kraig RP, Popko B (2008) Enhanced integrated stress response promotes myelinating oligodendrocyte survival in response to interferon-{gamma}. Am J Pathol 173:1508–1517

Lu PD, Harding HP, Ron D (2004a) Translation re-initiation at alternative open reading frames regulates gene expression in an integrated stress response. J Cell Biol 167:27–33

Lu PD, Jousse C, Marciniak SJ, Zhang Y, Novoa I, Scheuner D, Kaufman RJ, Ron D, Harding HP (2004b) Cytoprotection by pre-emptive conditional phosphorylation of translation initiation factor 2. EMBO J 23:169–179

Makioka K, Yamazaki T, Fujita Y, Takatama M, Nakazato Y, Okamoto K (2010) Involvement of endoplasmic reticulum stress defined by activated unfolded protein response in multiple system atrophy. J Neurol Sci 297:60–65

Marciniak SJ, Yun CY, Oyadomari S, Novoa I, Zhang Y, Jungreis R, Nagata K, Harding HP, Ron D (2004) Chop induces death by promoting protein synthesis and oxidation in the stressed endoplasmic reticulum. ' 18:3066–3077

Marciniak SJ, Garcia-Bonilla L, Hu J, Harding HP, Ron D (2006) Activation-dependent substrate recruitment by the eukaryotic translation initiation factor 2 kinase perk. J Cell Biol 172:201–209

Maurin A, Jousse C, Averous J, Parry L, Bruhat A, Cherasse Y, Zeng H, Zhang Y, Harding H, Ron D, Fafournoux P (2005) The gcn2 kinase biases feeding behavior to maintain amino-acid homeostasis in omnivores. Cell Metab 1:273–277

Mori K, Ma W, Gething MJ, Sambrook J (1993) A transmembrane protein with a cdc2+/cdc28-related kinase activity is required for signaling from the er to the nucleus. Cell 74:743–756

Murray JI, Whitfield ML, Trinklein ND, Myers RM, Brown PO, Botstein D (2004) Diverse and specific gene expression responses to stresses in cultured human cells. Mol Biol Cell 15:2361–2374

Nawrocki ST, Carew JS, Pino MS, Highshaw RA, Dunner K Jr, Huang P, Abbruzzese JL, McConkey DJ (2005) Bortezomib sensitizes pancreatic cancer cells to endoplasmic reticulum stress-mediated apoptosis. Cancer Res 65:11658–11666

Nishitoh H, Matsuzawa A, Tobiume K, Saegusa K, Takeda K, Inoue K, Hori S, Ishikawa K, Mizusawa H, Kakizuka A, Ichijo H (2002) Ask1 is essential for endoplasmic reticulum stress-induced neuronal cell death triggered by expanded polyglutamine repeats. Genes Dev 16:1345–1355

Novoa I, Zeng H, Harding H, Ron D (2001) Feedback inhibition of the unfolded protein response by gadd34-mediated dephosphorylation of eif2a. J Cell Biol 153:1011–1022

Novoa I, Zhang Y, Zeng H, Jungreis R, Harding HP, Ron D (2003) Stress-induced gene expression requires programmed recovery from translational repression. EMBO J 22:1180–1187

Obeng EA, Carlson LM, Gutman DM, Harrington WJ Jr, Lee KP, Boise LH (2006) Proteasome inhibitors induce a terminal unfolded protein response in multiple myeloma cells. Blood 107:4907–4916

Papa FR, Zhang C, Shokat K, Walter P (2003) Bypassing a kinase activity with an atp-competitive drug. Science 302:1533–1537

Pennuto M, Tinelli E, Malaguti MC, Del Carro U, Ron D, Quattrini A, Feltri ML, Wrabetz L (2008) Ablation of the upr-mediator chop restores motor function and reduces demyelination in charcot marie tooth 1b mice. Neuron 57:393–505

Ron D, Walter P (2007) Signal integration in the endoplasmic reticulum unfolded protein response. Nat Rev Mol Cell Biol 8:519–529

Ryu EJ, Harding HP, Angelastro JM, Vitolo OV, Ron D, Greene LA (2002) Endoplasmic reticulum stress and the unfolded protein response in cellular models of parkinson's disease. J Neurosci 22:10690–10698

Sado M, Yamasaki Y, Iwanaga T, Onaka Y, Ibuki T, Nishihara S, Mizuguchi H, Momota H, Kishibuchi R, Hashimoto T, Wada D, Kitagawa H, Watanabe TK (2009) Protective effect against Parkinson's disease-related insults through the activation of xbp1. Brain Res 1257:16–24

Scheuner D, Song B, McEwen E, Gillespie P, Saunders T, Bonner-Weir S, Kaufman RJ (2001) Translational control is required for the unfolded protein response and in vivo glucose homeostasis. Mol Cell 7:1165–1176

Schiffmann R, Elroy-Stein O (2006) Childhood ataxia with cns hypomyelination/vanishing white matter disease a common leukodystrophy caused by abnormal control of protein synthesis. Mol Genet Metab 88:7–15

Shamu CE, Walter P (1996) Oligomerization and phosphorylation of the ire1p kinase during intracellular signaling from the endoplasmic reticulum to the nucleus. EMBO J 15:3028–3039

Shen J, Chen X, Hendershot L, Prywes R (2002) Er stress regulation of atf6 localization by dissociation of bip/grp78 and unmasking of golgi localization signals. Dev Cell 3:99–111

Siekierka J, Mitsui KI, Ochoa S (1981) Mode of action of the heme-controlled translational inhibitor: relationship of eukaryotic initiation factor 2-stimulating protein to translation restoring factor. Proc Natl Acad Sci USA 78:220–223

Silva R, Ries V, Oo T, Yarygina O, Jackson-Lewis V, Ryu E, Lu P, Marciniak S, Ron D, Przedborski S, Greene L, Burke R (2005) Chop/gadd153 is a mediator of apoptotic death in substantia nigra dopamine neurons in an in vivo neurotoxin model of parkinsonism. J Neurochem 95:974–986

Smith WW, Jiang H, Pei Z, Tanaka Y, Morita H, Sawa A, Dawson VL, Dawson TM, Ross CA (2005) Endoplasmic reticulum stress and mitochondrial cell death pathways mediate a53t mutant alpha-synuclein-induced toxicity. Hum Mol Genet 14:3801–3811

Tamura T, Sunryd JC, Hebert DN (2010) Sorting things out through endoplasmic reticulum quality control. Mol Membr Biol 27:412–427

Travers KJ, Patil CK, Wodicka L, Lockhart DJ, Weissman JS, Walter P (2000) Functional and genomic analyses reveal an essential coordination between the unfolded protein response and er-associated degradation. Cell 101:249–258

Tribouillard-Tanvier D, Beringue V, Desban N, Gug F, Bach S, Voisset C, Galons H, Laude H, Vilette D, Blondel M (2008) Antihypertensive drug guanabenz is active in vivo against both yeast and mammalian prions. PLoS One 3:e1981

Tsaytler P, Harding HP, Ron D, Bertolotti A (2011) Selective inhibition of a regulatory subunit of protein phosphatase 1 restores proteostasis. Science 332:91–94

Uehara T, Nakamura T, Yao D, Shi ZQ, Gu Z, Ma Y, Masliah E, Nomura Y, Lipton SA (2006) S-nitrosylated protein-disulphide isomerase links protein misfolding to neurodegeneration. Nature 441:513–517

Urano F, Wang X, Bertolotti A, Zhang Y, Chung P, Harding H, Ron D (2000) Coupling of stress in the endoplasmic reticulum to activation of jnk protein kinases by transmembrane protein kinase ire1. Science 287:664–666

Vattem KM, Wek RC (2004) Reinitiation involving upstream orfs regulates atf4 mrna translation in mammalian cells. Proc Natl Acad Sci U S A 101:11269–11274

Wiseman RL, Zhang Y, Lee KP, Harding HP, Haynes CM, Price J, Sicheri F, Ron D (2010) Flavonol activation defines an unanticipated ligand-binding site in the kinase-rnase domain of ire1. Mol Cell 38:291–304

Yamamoto K, Sato T, Matsui T, Sato M, Okada T, Yoshida H, Harada A, Mori K (2007) Transcriptional induction of mammalian er quality control proteins is mediated by single or combined action of atf6alpha and xbp1. Dev Cell 13:365–376

Ye J, Rawson RB, Komuro R, Chen X, Dave UP, Prywes R, Brown MS, Goldstein JL (2000) Er stress induces cleavage of membrane-bound atf6 by the same proteases that process srebps. Mol Cell 6:1355–1364

Yoneda T, Imaizumi K, Oono K, Yui D, Gomi F, Katayama T, Tohyama M (2001) Activation of caspase-12, an endoplastic reticulum (er) resident caspase, through tumor necrosis factor receptor-associated factor 2-dependent mechanism in response to the er stress. J Biol Chem 276:13935–13940

Yoshida H, Haze K, Yanagi H, Yura T, Mori K (1998) Identification of the cis-acting endoplasmic reticulum stress response element responsible for transcriptional induction of mammalian glucose-regulated proteins. Involvement of basic leucine zipper transcription factors. J Biol Chem 273:33741–33749

Yoshida H, Matsui T, Yamamoto A, Okada T, Mori K (2001) Xbp1 mrna is induced by atf6 and spliced by ire1 in response to er stress to produce a highly active transcription factor. Cell 107:881–891

Zhang P, McGrath B, Li S, Frank A, Zambito F, Reinert J, Gannon M, Ma K, McNaughton K, Cavener DR (2002) The perk eukaryotic initiation factor 2 alpha kinase is required for the development of the skeletal system, postnatal growth, and the function and viability of the pancreas. Mol Cell Biol 22:3864–3874

Zhou D, Pallam LR, Jiang L, Narasimhan J, Staschke KA, Wek RC (2008) Phosphorylation of eif2 directs atf5 translational control in response to diverse stress conditions. J Biol Chem 283:7064–7073

Zinszner H, Kuroda M, Wang X, Batchvarova N, Lightfoot RT, Remotti H, Stevens JL, Ron D (1998) Chop is implicated in programmed cell death in response to impaired function of the endoplasmic reticulum. Genes Dev 12:982–995

Proteostasis and the aging pathways

Ian Nicastro and Andrew Dillin

Abstract Aging is defined as the continuous accumulation of changes in an organism over time, arising intrinsically as well as through interaction with the environment. Some of these changes are harmful to the maintenance of homeostasis and render an individual more susceptible to diseases and disorders. Eventually, accumulation of enough of these changes negatively impacts an organism's ability to respond to stresses and sets in motion a cascade of events that eventually ends in death. Aging is a universal condition that affects all life; however, different organisms age at remarkably different rates. This observation suggests that aging is not simply a stochastic response but instead has a powerful element of genetic mechanism. In seeking to understand why we age, there is potential for discovering strategies that might delay this inevitable process.

Introduction

Proteins exist in a dynamic and congested environment, one in which both intrinsic and extrinsic challenges often necessitate situations in which they must be rapidly synthesized or degraded. The stability of each protein must be carefully balanced between its capacity to remain folded and functional amid the challenges surrounding it and yet retain enough flexibility to allow for its degradation in order to achieve proper regulation of cellular functions. Even a slight change in energetic stability can result in proteins that rapidly transform into a misfolded or aggregated state. The majority of proteins have thus evolved within narrow thresholds of stabilities to ensure this dichotomy in regulation.

Unfortunately, the precariousness of this balance is dramatically exacerbated with age. As an organism ages, its proteins face an increasing severity in the

I. Nicastro • A. Dillin (✉)
Molecular and Cell Biology Laboratory, Salk Institute for Biological Studies, La Jolla, CA, USA
e-mail: dillin@salk.edu

R.I. Morimoto and Y. Christen (eds.), *Protein Quality Control in Neurodegenerative Diseases*, Research and Perspectives in Alzheimer's Disease,
DOI 10.1007/978-3-642-27928-7_3,

challenges they receive from environmental perturbations. The chaperones that a protein relies upon to fold become dysregulated while the machineries that are required for the protein's routine degradation stop working properly; the protein starts to misfold. Neighboring proteins likewise begin misfolding, aggregating, and sequestering away the remaining chaperones that could have been used to restore its proper form. At this point, the cell needs to mount a response to restore its homeostasis; however, the stress response machinery that it typically relies upon when faced with such challenges has lost its capacity to function. The collapse of the cell escalates.

This breakdown, however, does not immediately lead to complete entropy. As the organism ages, it exhibits a surprising degree of correlated, recognizable, and predictable changes to its physiology over time. These changes can occur synchronously across multiple tissues and organs. The phenotypic changes of aging occur in a type of concert, rather than in isolation, suggesting the residual participation of the endocrine system in the onset of age-related phenotypes. The demise of the cell thus most often occurs within the context of the simultaneous demise of the whole organism.

Over the past 30 years, researchers have elucidated a significant number of components of the signaling pathway that, after perturbation, significantly affect the longevity of the organism. Perhaps unsurprisingly, many of the genes that affect longevity also affect the coordination of whole-organism responses to the environment, such as metabolic pathways and endocrine signaling. Given the central role of protein homeostasis in the capacity of an organism to respond to stress and maintain overall cellular homeostasis, it seems probable that these aging pathways will also necessarily affect the regulation of proteostasis and subsequently the energetic stability of proteins within the cell. It is thus important to consider 1) what is known about the specific nature of the dysfunction and damage that affect proteins as the organism ages and 2) how the aging pathways mitigate this dysfunction.

What happens to proteins in an older individual?

Oxidative damage and carbonylation

One of the most widely examined theories of aging surrounds Denham Harman's Free Radical theory, which states that aging is the result of an accumulation of DNA damage from unpaired electron species known as free radicals. This theory has come to be known as the Oxidative Stress theory of aging because cellular metabolic processes generate reactive oxygen species (ROS), which cause oxidative damage to proteins, lipids and DNA over time (Cui et al. 2012). ROS include such molecules as superoxide anion, hydroxyl radical and hydrogen peroxide; the first two of these are free radicals whereas hydrogen peroxide is a potent oxidizer. ROS are primarily produced as a byproduct of mitochondrial respiration as electrons leak from the electron transport chain and react with oxygen.

Cells have multiple strategies for protecting themselves against oxidative damage, including multiple DNA repair mechanisms as well as antioxidants that scavenge ROS and neutralize them. These defense strategies become weakened as an organism ages and eventually are overwhelmed. Reduced free radical production with increased longevity as well as observations of shortened life span in animals lacking antioxidant genes (Elchuri et al. 2005; Finkel 2000) have been described. Additionally, reduced expression of genes that encode for components of the electron transport chain has been shown to extend life span in invertebrate model systems (Dillin et al. 2002b; Feng et al. 2001; Lee et al. 2002).

To the aging protein, an attack by free radicals can result in oxidative damage that results in its differential targeting for degradation. Oxidative damage can only be re-reduced on two of the 20 amino acids, rendering most damaged proteins as difficult to repair (Mary et al. 2004). The presence of oxidative damage often causes changes to the utility of the proteins, frequently targeting stretches of polypeptides that are of functional import. Often times, oxidative damage results in the specific formation of reactive ketones or aldehydes, also called protein carbonylation. Protein carbonylation is especially problematic for the protein, as it is both irreparable and irreversible (Nystrom 2005). Over the years, carbonylation in itself has become a widely used measure for overall levels of oxidative stress (Suzuki et al. 2010).

As the oxidative damage on the protein increases, its propensity to misfold also increases. While moderately oxidized proteins are targeted for degradation by the proteasome, heavy levels of oxidation can eventually lead to crosslinking with other proteins to an extent at which the protein can no longer be readily degraded by the proteasome (Jung et al. 2007). Thus, perturbations that affect either ROS production, ROS scavenging, or proteasome function can all play a role in the eventual accumulation of damaged proteins within the cell.

Loss of chaperone capacity

In in vivo settings, proteins rarely can fold or retain their folded states in the absence of chaperones. Within the crowded cell, which can face proteins in concentrations in excess of 200 mg/ml, a decrease in the availability of chaperones almost universally has a negative effect on the stability or appropriate regulation of proteins. For the aging organism, the coordination of an appropriate number of chaperones, even as it faces a rapidly increasingly level of damage in its pool of aging proteins, remains a daunting task.

Many of the chaperones are regulated through stress response pathways that originated to protect the cell. Chaperones derive their nomenclature based on their apparent molecular weight. For example, HSP70 and HSP90 are high molecular weight chaperones that bind unfolded peptide sequences and form complexes with co-chaperones to refold target proteins in an ATP-dependent manner (Lanneau et al. 2010). In contrast, HSP10 and HSP60 are ATP-dependent chaperones that belong to the chaperonin family and oligomerize to create barrel-shaped structures that catalyze protein refolding within (Calderwood et al. 2009). Smaller molecular weight chaperones, such as HSP27 and the *C. elegans*-specific HSP16, comprise the small heat shock protein (sHSP) family, whose members undergo oligomerization to form dynamic higher molecular weight complexes that refold target proteins in an ATP-independent manner (Lanneau et al. 2010).

The predominant regulation of chaperones used in the folding of proteins is coordinated by the cytoplasmic heat responsive transcription factor, HSF-1. HSF-1 induces transcription of multiple families of HSPs, catalyzing the unfolding and subsequent proper re-folding of misfolded proteins (Calderwood et al. 2009). HSF-1 and the heat shock response play a role in both aging and age-associated diseases involving aggregation-prone peptides, such as Huntington's and Alzheimer's diseases. The functionality of the heat shock response has been shown to decline significantly with age in *C. elegans* and mammals (Ben-Zvi et al. 2009; Calderwood et al. 2009).

Conversely, elevated expression of specific HSPs (HSP16 and HSP90) has been observed in long-lived *C. elegans*, and overexpression of HSF-1, HSP70 or HSP16 has been shown to produce life span extension (Hsu et al. 2003; Morley and Morimoto 2004; Walker and Lithgow 2003; Yokoyama et al. 2002). Down regulation of *hsf-1* dramatically reduces life span in *C. elegans* and results in an accelerated onset of polyglutamine aggregation in a transgenic *C. elegans* model of Huntington's disease (Garigan et al. 2002; Hsu et al. 2003; Morley and Morimoto 2004).

Cellular stresses, such as exposure to heat, toxins or oxidative stress, increase the probability that proteins will undergo misfolding and be rendered unable to function properly (Lindquist and Craig 1988). Additionally, misfolded proteins that possess exposed hydrophobic regions can undergo aggregation and cause cellular toxicity by disrupting the function of other proteins as well as membranes (Stefani and Dobson 2003). It is therefore critical that protein folding is maintained during times of stress, and the heat shock response pathway is the primary mechanism by which this is achieved. Heat shock factor (HSF) is a transcription factor that rapidly induces expression of heat shock proteins (HSPs) in response to cellular stress in almost every prokaryotic and eukaryotic organism (Anckar and Sistonen 2011). Mammals have three HSFs (HSF1, HSF2 and HSF4) with HSF1 acting in a stress-inducible fashion, whereas invertebrates such as *C. elegans* possess only one HSF, HSF-1 (Taylor and Dillin 2011).

HSF-1 is arguably one of the most important regulators of proteostasis in the cell. During non-stressed conditions, HSF-1 remains as an inactive monomer in a complex with HSP90 and additional cofactors that release HSF-1 upon exposure to stress, allowing it to undergo homo-trimerization to its transcriptionally active form (Anckar and Sistonen 2011). *hsf-1* knockdown in a *C. elegans* model of Alzheimer's disease dramatically accelerated the rate of observed proteotoxicity and led to increased accumulation of both small and high molecular weight Aβ aggregates (Cohen et al. 2006). *hsf-1* appears to regulate genes that act to disaggregate and degrade highly toxic Aβ aggregates, and it is believed that this pathway operates preferentially to a secondary active aggregation pathway regulated by

DAF-16 (Cohen et al. 2006). Additional findings have shown that *hsf-1* needs to be expressed during larval development to orchestrate this protective response against Aβ (Cohen et al. 2006; Volovik et al. 2012). However, it is likely that a loss of HSF-1 activity diminishes the cell's capacity to respond to proteotoxic stress at older ages.

Changes in the regulation of autophagy

Autophagy is the recycling of organelles and cytoplasmic constituents through a process that involves sequestering and then delivering these materials to lysosomes for degradation (Yang and Klionsky 2009). Different avenues of acquisition and subsequent delivery to lysosomes exist, including macroautophagy, chaperone-mediated autophagy and microautophagy. Macroautophagy is well conserved among eukaryotes and is the best characterized of the three avenues of autophagy, with the term autophagy often used to refer specifically to macroautophagy (Mizushima et al. 2010). Macroautophagy begins with the formation of an isolation membrane that elongates and eventually fuses on itself to surround a portion of cytoplasm, forming a double-membraned vesicle termed an autophagosome (Rubinsztein et al. 2011). The outer membrane of the autophagosome fuses with a lysosome to form an autolysosome, which leads to degradation of the inner membrane and the cytoplasmic contents within it by lysosomal hydrolases (proteases, lipases, glycosideases, and nucleotidases). The resulting amino acids and other molecules liberated by hydrolysis are then returned to the cytoplasm through permeases for recycling (Rubinsztein et al. 2011).

Autophagy is highly regulated and it functions at relatively low levels until specific cellular stresses are detected (Rubinsztein et al. 2011). Starvation is one of the most potent inducers of the autophagy response, as breakdown of cytoplasmic macromolecules can serve as an emergency source of energy. Autophagy has been shown to be induced under metabolic perturbations that extend longevity, such as dietary restriction (DR), and the longevity associated with DR requires specific genetic components of the autophagy pathway (Hansen et al. 2008). The highly conserved eukaryotic nutrient sensor TOR (Target of Rapamycin) also plays a well-characterized regulatory role by which it inhibits macroautophagy under normal nutrient conditions (Suzuki and Ohsumi 2007). Atg1 is a component of the Atg1/unc-51-like kinase (ULK) complex required for initiation of the isolation membrane, and TOR inhibits Atg1 activity when a cell's energy needs are met (Yang and Klionsky 2009). Under low nutrient conditions, TOR inhibition of Atg1 is removed and ULK activity leads to formation of a protein complex that includes the class III PI3K Vps34 and Beclin1 (*bec-1* in *C. elegans*), which both also serve a necessary role in initiation of the isolation membrane (Ravikumar et al. 2010). Approximately 35 genes have been shown to function in macroautophagy, most of which are identified by the nomenclature prefix Atg. Atg8, also known as LC3 in mammals and *lgg-1* in *C. elegans*, serves a necessary role in the elongation of autophagosome membranes and is a commonly used marker for autophagosomes

(Mizushima et al. 2010). Another important autophagy gene is Atg18, atg-18 in *C. elegans*, which possesses WD40 domains and is an effector protein for the PI3K enzymatic product PI3P (Lu et al. 2011).

Chaperone-mediated autophagy involves the use of chaperones to bind and deliver specific soluble cytosolic proteins to lysosomes for degradation (Wong and Cuervo 2010). The chaperone hsc70 is a constitutively expressed hsp70 family member that recognizes specific motifs in target proteins and complexes with them (Li et al. 2011a). Hsc70 and its target bind Lamp2a receptors on a lysosomal membrane, and the target protein is unfolded and then translocated into the lysosomal lumen for degradation (Li et al. 2011a). The heat shock response pathway chaperones hsp90 and hsp40, as well as CHIP, act as cofactors for hsc70 and regulate its activity (Li et al. 2011a). While these components of the heat shock response pathway have only been shown to play a direct role in chaperone-mediated autophagy, indirect evidence indicates that *hsf-1* may also function in macroautophagy. This evidence hinges on the fact that DR and IIS, which are dependent on *hsf-1*, both activate macroautophagy (Hansen et al. 2008).

Autophagy has been indicated in the degradation of aggregation-prone peptides associated with neurodegenerative diseases such as Parkinson's, Huntington's and Alzheimer's (Menzies et al. 2011). Both Aβ-expressing *C. elegans* and the brains of Alzheimer's disease patients exhibit accumulations of autophagic vacuoles that have failed to proceed with lysosomal fusion, suggesting that Aβ impairs autophagy (Boland et al. 2008; Florez-McClure et al. 2007). Human Alzheimer's disease patients exhibit reduced levels of Beclin1, and genetic knockdown of Beclin1 in the brains of Alzheimer's disease model mice accelerated neurodegeneration and Aβ accumulation (Pickford et al. 2008).

Evidence has shown that ubiquitination can mark misfolded and aggregated proteins for removal by a recently discovered form of receptor-mediated macroautophagy (Tyedmers et al. 2010). This process acts as a link between the ubiquitin proteasome degradation system and autophagy, yielding an additional avenue besides chaperone-mediated autophagy in which heat shock response proteins can direct targets for autophagic degradation. This collection of observations suggests a significant role for autophagy in the removal of aggregation-prone peptides associated with neurodegenerative disease and other age-related diseases. However, the extent to which autophagy is differentially regulated in an aging individual remains somewhat unclear. Without the proper function of autophagy pathways, however, the accumulation of damaged proteins can quickly become overwhelming to the organism. Thus, the relationship between autophagy activation and health span remains essential.

Changes in the ubiquitin proteasome system

The ubiquitin proteasome system is a well-conserved pathway that plays a significant role in the degradation of proteins that reside in the cytosol, nucleus and endoplasmic reticulum (ER) of eukaryotes. Proteins are signaled for degradation by the proteasome through ubiquitination, a process by which the small globular peptide ubiquitin is conjugated to a target protein in succession to form a small ubiquitin polymer (Pickart 2001). Ubiquitination does not always result in proteasome degradation of the target, as some ubiquitin linkages serve a signaling role or modify the targets function (Wong and Cuervo 2010). Ubiquitin conjugation employs a E1 ubiquitin activating enzyme, a E2 ubiquitin conjugating enzyme, and a E3 ubiquitin ligase (Scheffner et al. 1995). E3 ubiquitin ligases are responsible for the selection of protein substrate, usually through recognition and binding of exposed degradation sequences found on unstable proteins. Eukaryotes possess over 1,000 E3s that allow for significant specificity in targeting proteins for degradation by the proteasome (Staub and Rotin 2006).

The proteasome is a large protease complex consisting of a catalytic 20S core segment and an 11S or 19S regulatory cap that can dock at one or both sides of the core to form proteasome species of various sizes (Wong and Cuervo 2010). The 26S proteasome, which is formed by one 19S cap and a 20S core, plays the most well-characterized role in protein degradation and quality control. The 20S core is comprised of two outer α rings that serve as a gateway to the two inner β rings that function as the site of proteolysis (Pickart and Cohen 2004). Ubiquitinated protein substrates that arrive at the regulatory cap of a 26S proteasome undergo deubiquitination and unfolding, which requires energy supplied by ATPase subunits present in the regulatory cap. The target peptides are then passed through the α ring gateway and exposed to the potent caspase-like, trypsin-like, and chymotrypsin-like protease activities of the β rings (Wong and Cuervo 2010).

The transcription factor *hsf-1* serves an indirect regulatory role in the ubiquitin proteasome system. HSP targets of *hsf-1* act as E3 cofactors and interact directly with assembled proteasomes. For example, CHIP is an important E3 that interacts with members of the hsp70 and hsp90 families of heat shock pathway chaperones to ubiquitinate target proteins and flag them for destruction by the proteasome (Kumar et al. 2007; Tyedmers et al. 2010). CHIP and hsp70 have been shown to function in mechanisms that reduce Aβ levels in mammalian systems (Kumar et al. 2007). hsp90 also plays a critical role in maintaining the structure and function of fully assembled proteasome complexes (Imai et al. 2003). *skn-1* plays an even more direct regulatory role in proteasome function, as it serves as a transcription factor for specific subunits of the proteasome (Kahn et al. 2008; Li et al. 2011b). The transcription factor *skn-1*, which responds to increases in oxidative stress such as those seen with age, also activates *aip-1* (arsenic inducible RNA-associated protein), which is known to bind to and enhance function of the proteasome as well as contribute to degradation of Aβ (Ferguson et al. 2010; Hassan et al. 2009).

Human Alzheimer's disease patients exhibit a significant decrease in proteasome activity in specific brain regions compared to healthy controls (Riederer et al. 2011). Additionally, colocalization of 26S proteasomes with protein aggregates as well as observations of delayed removal of aggregates in the presence of proteasome inhibition yield further evidence for the role of proteasome degradation of toxic peptides (Tyedmers et al. 2010).

Aging pathways: the genetics behind the proteostasis of a cell

The importance for proteostasis in the health of the cell suggests that genetic manipulations that directly influence aging will also influence the proteostasis of the cell. Studies of aging in the nematode *C. elegans* have helped identify components of several genetic pathways that regulate aging, including insulin signaling, dietary restriction, and mitochondrial respiration.

Insulin/IGF-1 signaling and proteostasis

Research conducted in both invertebrate and mammalian model systems has shown that perturbations in the insulin/insulin-like growth factor signaling (IIS) pathway result in profound extension of longevity (Bartke 2008). The genetic components of the IIS pathway are highly conserved between mammals and invertebrates, which has led to *C. elegans* being extensively used to explore this pathway. In *C. elegans,* the IIS pathway has one insulin/IGF-1 receptor, DAF-2, which, when bound by insulin-like ligands, activates the phosphatidylinositol 3-kinase AGE-1 to produce PI3 (Morris et al. 1996). The phosphatase DAF-18, homologous to mammalian PTEN, functions as an 'off switch' to oppose the activity of AGE-1 (Ogg et al. 1997). PI3 produced by AGE-1 activates PDK-1 kinase, which in turn phosphorylates the kinases AKT-1, AKT-2 and SGK-1 (Hertweck et al. 2004; Paradis et al. 1999; Paradis and Ruvkun 1998). This kinase trio phosphorylates the forkhead transcription factor DAF-16 (orthologous to FOXO1, FOXO3a, and FOXO4 in mammals), causing it to associate with the 14-3-3 protein FTT-2 in the cytosol that keeps it excluded from gene targets in the nucleus (Cahill et al. 2001; Li et al. 2007).

Upon a reduction in insulin signaling, un-phosphorylated DAF-16 enters the nucleus, where it associates with co-activators like SMK-1 and binds to an evolutionarily conserved consensus sequence (Wolff et al. 2006). Additional factors function to antagonize DAF-16 within the nucleus, such as host cell factor 1 (HCF-1), which can bind DAF-16 and prevent it from interacting with target genes (Li et al. 2008). DAF-16 activates a broad range of genes that lead to increased longevity as well as amplified resistance against heat and oxidative damage (Lin et al. 1997; Ogg et al. 1997), including genes that neutralize ROS like the superoxide dismutase *sod-3 and the* catalases *ctl-1* and *ctl-2,* as well as genes involved in metabolism and the heat shock response (McElwee et al. 2004; Murphy et al. 2003). Knocking down any single transcriptional target of DAF-16 fails to fully prevent IIS-mediated longevity, revealing just how complicated the life span response is (McElwee et al. 2004; Murphy et al. 2003).

DAF-16 is not the only factor regulated by DAF-2 and the IIS pathway, as two other transcription factors serve roles in IIS-mediated longevity. HSF-1, which is known for orchestrating the cellular heat shock response, has been shown to be essential for IIS pathway-induced life span extension by DAF-16 (Hsu et al. 2003; Morley and Morimoto 2004). Despite this observation, the exact signaling mechanism by which DAF-2 regulates HSF-1 remains to be elucidated. The additional transcription factor regulated by IIS is SKN-1, which serves a previously characterized role as the master regulator of the oxidative stress response and has been shown to be directly inhibited by the kinase cascade downstream of DAF-2 (Tullet et al. 2008). SKN-1 is not required for IIS-induced longevity but does appear necessary to obtain the maximal life span extension possible from modulating IIS. Neither HSF-1 nor SKN-1 is considered a canonical member of the IIS pathway due to their previously characterized roles in regulating other important pathways. The regulation of HSF-1 and SKN-1 by DAF-2 demonstrates that DAF-16 is not solely responsible for IIS-mediated longevity, and it is possible that other factors remain to be discovered.

Aside from its effects on aging, reduced IIS confers a significant benefit in the proteostasis of the cell. Reduced *daf-2* also activates the transcription factor HSF-1, subsequently upregulating many of the cytosolic heat shock chaperones responsible for proteostasis (Chiang et al. 2012; Garigan et al. 2002; Hsu et al. 2003; Morley and Morimoto 2004). Reduced IGF-1/insulin signaling also protects in a large number of proteotoxic, age-onset neurodegenerative disease models, including those tested in mice (Brignull et al. 2006; Cohen et al. 2006, 2009; Dillin and Cohen 2011).

Dietary restriction and proteostasis

A significant reduction in dietary caloric intake, known as DR, dramatically increases the life span of a wide assortment of eukaryotic organisms (Weindruch et al. 1988). DR is generally defined as a 30-60 % decrease in calorie intake compared to ad libitum feeding, without leading to malnutrition of the organism. It is important to point out that DR is not simply a process that removes harmful overfeeding conditions but rather an actual significant reduction below an organism's baseline calorie intake. DR produces a graded parabolic response, where the progressive reduction of calories results in increasing longevity until a level of calorie reduction is achieved that yields the greatest increase in life span (Mair and Dillin 2008). Reducing calories beyond that point will result in progressively less optimal life span extension as malnutrition begins to set in. At levels of extreme caloric reduction, malnutrition is so significant that little or no life span extension is observed. Protocols for applying DR vary widely, with some focusing on decreasing levels of a specific dietary macronutrient like protein whereas others represent a cutback in all dietary macronutrients (Houthoofd and Vanfleteren 2006). The end result of these varied protocols is still the same: a total reduction in overall caloric intake without malnutrition that leads to increased longevity.

Remarkably, DR appears to be able to increase longevity in almost every eukaryotic species so far tested, suggesting that it is a public mechanism of life span extension. DR has been shown to extend the life span of the classical model organisms: the yeast *Saccharomyces cerevisiae*, the nematode *Caenorhabditis elegans*, the fruit fly *Drosophila melanogaster*, the mouse *Mus musculus*, and the rat *Rattus norvegicus* (Bishop and Guarente 2007a). Longevity extension has also been observed for less commonly studied animals, including insects, fish and dogs placed on DR regimes (Mair and Dillin 2008). Several long-term studies were initiated to examine the effects of DR on primates; however, these experiments are still ongoing. The initial observations from these primate studies have demonstrated that physiological changes associated with aging are in fact delayed in the DR groups (Rezzi et al. 2009). Additionally, the preliminary life span data from the few monkeys that have died so far are suggestive that DR may also extend life span in primates (Colman et al. 2009). Large-scale experimentation of humans adhering to DR diet regimes has only recently been undertaken, and so the effects of DR on human longevity remain to be seen (Holloszy and Fontana 2007).

Importantly, DR invokes the initiation of a signaling pathway that can be genetically dissected and manipulated. Studies using the nematode *C. elegans* have implicated four genes, *hsf-1, skn-1, pha-4,* and *wwp-1*, as being required for DR-mediated life span extension (Bishop and Guarente 2007b; Carrano et al. 2009; Panowski et al. 2007; Steinkraus et al. 2008). Each of these genes has a previously characterized vital role in physiology: *hsf-1* orchestrates the cellular heat shock response, *skn-1* regulates the oxidative stress pathway, *pha-4* is required for foregut development, and *wwp-1* is an E3 ligase necessary for embryonic development (An and Blackwell 2003; Carrano et al. 2009; Horner et al. 1998; Hsu et al. 2003). Three of these, *hsf-1, skn-1,* and *pha-4,* are transcription factors with well-characterized transcriptional targets. HSF-1 activates expression of molecular chaperones known as HSPs, SKN-1 targets a broad array of detoxification and antioxidant genes, and PHA-4 induces expression of ROS scavenging superoxide dismutases (Hsu et al. 2003; Panowski et al. 2007; Park et al. 2009). Interestingly, there is some overlap between the transcriptional profiles for these factors; for example, PHA-4 and SKN-1 both activate expression of *sod-1*, a cytosolic Cu/Zn superoxide dismutase (Panowski et al. 2007; Park et al. 2009). There are five sod genes in *C. elegans*, none of which have been shown to be single-handedly required for DR induced longevity. In fact, there is notable redundancy in function for many targets of all three of these transcription factors, suggesting that the longevity obtained from DR involves a complicated multi-gene downstream response.

Beyond simply making an organism live longer, DR also appears to prolong a youthful state by delaying the onset of many diseases associated with old age. The state of proteostasis within an organism is directly perturbed when that organism is placed under conditions of DR. Thus far, of the four genes that have been shown to be required for DR-induced life span extension, only *hsf-1* has been tested to determine its role in the DR-mediated protective response (Steinkraus et al. 2008). *hsf-1* was shown to be required for to extend life span and protect against observed proteotoxicity in *C. elegans* models of Huntington's and Alzheimer's diseases, and this role of *hsf-1* was demonstrated to be independent of IIS function (Steinkraus et al. 2008). These results were interpreted to suggest that *hsf-1* is required for DR-induced life span extension as well as DR-mediated protection from proteotoxicity.

Mitochondrial function and proteostasis

In a number of model organisms, mutations in genes that affect mitochondria function have been found that dramatically increase the organism's life span. Work in *C. elegans* suggests that the relative state of electron transport chain (ETC.) function of the mitochondria at a specific stage in life can be signaled and perpetuated to set a rate of aging for the organism even after ETC. function is restored (Bratic et al. 2009; Dillin et al. 2002b; Mun et al. 2010). In part because the mitochondria serve as the primary source of energy for the cell, theories of aging have repeatedly pointed towards a compelling correlation between the health of the mitochondria and the rate at which an organism ages. The mitochondrial theory of aging originated with the idea that mutations in mitochondrial DNA (mtDNA) accumulate as the organism ages (Harman 1972). A loss of fidelity in mtDNA maintenance creates a catastrophic, positive feedback loop in which a poorly functioning mitochondrion both diminishes its capacity for ATP production and increases its ROS production. ROS damage to the macromolecular complexes and DNA within the mitochondria causes even more ROS production, and the mitochondrion enters into a vicious cycle in which it accelerates its own demise, leading toward the eventual senescence or apoptosis of the whole cell. Correlatively, tissues that depend upon respiration for their function, such as the neurons, skeletal muscles, and endocrine tissues, exhibit heightened vulnerability to mitochondrial DNA mutation or disease (Chan 2006). How, then, does the aging of a mitochondrion affect proteostasis in the entire organism?

Recent reports have described a role for the mitochondrial unfolded protein response (UPR^{mt}) in the regulation of longevity. In this study, downstream transcription factors required for the mitochondrial unfolded protein response specifically block the increased longevity in mitochondrial mutants but do not affect long-lived organisms originating from reduced insulin signaling or DR (Durieux et al. 2011), suggesting that proteostasis of the organism, as sensed and regulated by the mitochondria, has the capacity to affect overall longevity. The essential role that the mitochondrion has in cellular homeostasis and energy production suggests that it may act as a reactive sensor of random intrinsic or extrinsic variables capable of influencing an organism's susceptibility to disease. Changes within the mitochondria thus also might be responsible for the emergent properties displayed in such a system in response to stochastic changes and/or may play a significant role in coordinating the activation of non-mitochondrial stress response pathways, and especially those involved with proteostasis.

Studies have shown that mitochondria-mediated longevity functions genetically independent of other pathways known to enhance longevity, such as perturbations in the IIS signaling pathway or DR (Wolff and Dillin 2006). The independence of these three pathways arises from the fact that each possesses different temporal requirements as well as unique genetic components. Mitochondrial respiration must be downregulated during early development to extend life span, whereas reduced IIS signaling needs to take place during early to mid adulthood to result in increased longevity (Dillin et al. 2002a, b). Unlike the other two pathways, DR is able to extend longevity when imposed at any point during an organism's life span (Mair and Dillin 2008). Each pathway has key genetic components that are dispensable for the longevity associated with modulating other pathways; genetic components of the ETC. are unique to the mitochondrial respiration pathway, *daf-16* is unique to the IIS pathway, and DR has unique transcription factors (Dillin et al. 2002a; Houthoofd et al. 2003; Panowski et al. 2007; Steinkraus et al. 2008). DR-treated organisms do not show a reduction in metabolic rate, providing additional evidence for independence between the DR and mitochondrial respiration pathways (Houthoofd et al. 2002). Lastly, modulation of more than one pathway produces a super-additive on longevity, which further suggests that each pathway extends life span through an independent mechanism (Dillin et al. 2002a; Lakowski and Hekimi 1998).

Conclusion

The establishment and maintenance of homeostasis are arguably two of the most basic and yet intricately complex needs of an organism. In model organisms, the manipulation of each of the genetic pathways that affects the aging of the whole organism also drastically affects its capacity for maintaining proteostasis in the face of the challenges of old age. Further exploration of the required genetic components of these pathways and specific effectors of proteostasis may continue to yield valuable insights into the origins of age-onset proteotoxic diseases.

References

An JH, Blackwell TK (2003) SKN-1 links C. elegans mesendodermal specification to a conserved oxidative stress response. Genes Dev 17:1882–1893

Anckar J, Sistonen L (2011) Regulation of HSF1 function in the heat stress response: implications in aging disease. Ann Rev Biochem 80:1089–1115

Bartke A (2008) Insulin and aging. Cell Cycle 7:3338–3343

Ben-Zvi A, Miller EA, Morimoto RI (2009) Collapse of proteostasis represents an early molecular event in Caenorhabditis elegans aging. Proc Natl Acad Sci USA 106:14914–14919

Bishop NA, Guarente L (2007a) Genetic links between diet lifespan: shared mechanisms from yeast to humans. Nat Rev Genet 8:835–844

Bishop NA, Guarente L (2007b) Two neurons mediate diet-restriction-induced longevity in C. elegans. Nature 447:545–549

Boland B, Kumar A, Lee S, Platt FM, Wegiel J, Yu WH, Nixon RA (2008) Autophagy induction and autophagosome clearance in neurons: relationship to autophagic pathology in Alzheimer's disease. J Neurosci 28:6926–6937

Bratic I, Hench J, Henriksson J, Antebi A, Burglin TR, Trifunovic A (2009) Mitochondrial DNA level but not active replicase is essential for Caenorhabditis elegans development. Nucl Acids Res 37:1817–1828
Brignull HR, Moore FE, Tang SJ, Morimoto RI (2006) Polyglutamine proteins at the pathogenic threshold display neuron-specific aggregation in a pan-neuronal Caenorhabditis elegans model. J Neurosci 26:7597–7606
Cahill CM, Tzivion G, Nasrin N, Ogg S, Dore J, Ruvkun G, Alexer-Bridges M (2001) Phosphatidylinositol 3-kinase signaling inhibits DAF-16 DNA binding function via 14-3-3-dependent 14-3-3-independent pathways. J Biol Chem 276:13402–13410
Calderwood SK, Murshid A, Prince T (2009) The shock of aging: molecular chaperones the heat shock response in longevity and aging – a mini-review. Gerontology 55:550–558
Carrano AC, Liu Z, Dillin A, Hunter T (2009) A conserved ubiquitination pathway determines longevity in response to diet restriction. Nature 460:396–399
Chan DC (2006) Mitochondria: dynamic organelles in disease, aging and development. Cell 125:1241–1252
Chiang WC, Ching TT, Lee HC, Mousigian C, Hsu AL (2012) HSF-1 regulators DDL-1/2 link insulin-like signaling to heat-shock responses and modulation of longevity. Cell 148:322–334
Cohen E, Bieschke J, Perciavalle RM, Kelly JW, Dillin A (2006) Opposing activities protect against age-onset proteotoxicity. Science 313:1604–1610
Cohen E, Paulsson JF, Blinder P, Burstyn-Cohen T, Du D, Estepa G, Adame A, Pham HM, Holzenberger M, Kelly JW, Masliah E, Dillin A (2009) Reduced IGF-1 signaling delays age-associated proteotoxicity in mice. Cell 139:1157–1169
Colman RJ, Anderson RM, Johnson SC, Kastman EK, Kosmatka KJ, Beasley TM, Allison DB, Cruzen C, Simmons HA, Kemnitz JW, Weindruch R (2009) Caloric restriction delays disease onset and mortality in rhesus monkeys. Science 325:201–204
Cui H, Kong Y, Zhang H (2012) Oxidative stress mitochondrial dysfunction and aging. J Signal Transduct. doi:10.1155/2012/646354
Dillin A, Cohen E (2011) Ageing protein aggregation-mediated disorders: from invertebrates to mammals. Philos Trans R Soc Lond B Biol Sci 366:94–98
Dillin A, Crawford DK, Kenyon C (2002a) Timing requirements for insulin/IGF-1 signaling in C. elegans. Science 298:830–834
Dillin A, Hsu A-L, Arantes-Oliveira N, Lehrer-Graiwer J, Hsin H, Fraser AG, Kamath RS, Ahringer J, Kenyon C (2002b) Rates of behavior aging specified by mitochondrial function during development. Science 298:2398–2401
Durieux J, Wolff S, Dillin A (2011) The cell-non-autonomous nature of electron transport chain-mediated longevity. Cell 144:79–91
Elchuri S, Oberley TD, Qi W, Eisenstein RS, Jackson Roberts L, Van Remmen H, Epstein CJ, Huang TT (2005) CuZnSOD deficiency leads to persistent and widespread oxidative damage and hepatocarcinogenesis later in life. Oncogene 24:367–380
Feng J, BussiËre F, Hekimi S (2001) Mitochondrial electron transport is a key determinant of life span in Caenorhabditis elegans. Dev Cell 1:633–644
Ferguson AA, Springer MG, Fisher AL (2010) skn-1-Dependent and -independent regulation of aip-1 expression following metabolic stress in Caenorhabditis elegans. Mol Cell Biol 30:2651–2667
Finkel T (2000) Redox-dependent signal transduction. FEBS Lett 476:52–54
Florez-McClure ML, Hohsfield LA, Fonte G, Bealor MT, Link CD (2007) Decreased insulin-receptor signaling promotes the autophagic degradation of beta-amyloid peptide in C. elegans. Autophagy 3:569–580
Garigan D, Hsu AL, Fraser AG, Kamath RS, Ahringer J, Kenyon C (2002) Genetic analysis of tissue aging in Caenorhabditis elegans: a role for heat-shock factor and bacterial proliferation. Genetics 161:1101–1112
Hansen M, Chra A, Mitic LL, Onken B, Driscoll M, Kenyon C (2008) A role for autophagy in the extension of lifespan by dietary restriction in C. elegans. PLoS Genet 4:e24
Harman D (1972) The biologic clock: the mitochondria? J Am Geriatr Soc 20:145–147

Hassan WM, Merin DA, Fonte V, Link CD (2009) AIP-1 ameliorates beta-amyloid peptide toxicity in a Caenorhabditis elegans Alzheimer's disease model. Hum Mol Genet 18:2739–2747
Hertweck M, Gobel C, Baumeister R (2004) C elegans SGK-1 is the critical component in the Akt/PKB kinase complex to control stress response and life span. Dev Cell 6:577–588
Holloszy JO, Fontana L (2007) Caloric restriction in humans. Exp Gerontol 42:709–712
Horner MA, Quintin S, Domeier ME, Kimble J, Labouesse M, Mango SE (1998) pha-4 an HNF-3 homolog specifies pharyngeal organ identity in Caenorhabditis elegans. Genes Dev 12:1947–1952
Houthoofd K, Vanfleteren JR (2006) The longevity effect of dietary restriction in Caenorhabditis elegans. Exp Gerontol 41:1026–1031
Houthoofd K, Braeckman BP, Lenaerts I, Brys K, De Vreese A, Van Eygen S, Vanfleteren JR (2002) No reduction of metabolic rate in food restricted Caenorhabditis elegans. Exp Gerontol 37:1359–1369
Houthoofd K, Braeckman BP, Johnson TE, Vanfleteren JR (2003) Life extension via dietary restriction is independent of the Ins/IGF-1 signalling pathway in Caenorhabditis elegans. Exp Gerontol 38:947–954
Hsu AL, Murphy CT, Kenyon C (2003) Regulation of aging age-related disease by DAF-16 and heat-shock factor. Science 300:1142–1145
Imai J, Maruya M, Yashiroda H, Yahara I, Tanaka K (2003) The molecular chaperone Hsp90 plays a role in the assembly and maintenance of the 26S proteasome. Embo J 22:3557–3567
Jung T, Bader N, Grune T (2007) Oxidized proteins: intracellular distribution and recognition by the proteasome. Arch Biochem Biophys 462:231–237
Kahn NW, Rea SL, Moyle S, Kell A, Johnson TE (2008) Proteasomal dysfunction activates the transcription factor SKN-1 and produces a selective oxidative-stress response in Caenorhabditis elegans. Biochem J 409:205–213
Kumar P, Ambasta RK, Veereshwarayya V, Rosen KM, Kosik KS, Band H, Mestril R, Patterson C, Querfurth HW (2007) CHIP and HSPs interact with beta-APP in a proteasome-dependent manner and influence Abeta metabolism. Hum Mol Genet 16:848–864
Lakowski B, Hekimi S (1998) The genetics of caloric restriction in Caenorhabditis elegans. Proc Natl Acad Sci USA 95:13091–13096
Lanneau D, Wettstein G, Bonniaud P, Garrido C (2010) Heat shock proteins: cell protection through protein triage. Sci World J 10:1543–1552
Lee SS, Lee RYN, Fraser AG, Kamath RS, Ahringer J, Ruvkun G (2002) A systematic RNAi screen identifies a critical role for mitochondria in C. elegans longevity. Nat Genet 33:40–48
Li J, Tewari M, Vidal M, Lee SS (2007) The 14-3-3 protein FTT-2 regulates DAF-16 in Caenorhabditis elegans. Dev Biol 301:82–91
Li J, Ebata A, Dong Y, Rizki G, Iwata T, Lee SS (2008) Caenorhabditis elegans HCF-1 functions in longevity maintenance as a DAF-16 regulator. PLoS Biol 6:e233
Li W, Yang Q, Mao Z (2011a) Chaperone-mediated autophagy: machinery regulation biological consequences. Cell Mol Life Sci 68:749–763
Li X, Matilainen O, Jin C, Glover-Cutter KM, Holmberg CI, Blackwell TK (2011b) Specific SKN-1/Nrf stress responses to perturbations in translation elongation proteasome activity. PLoS Genet 7:e1002119
Lin K, Dorman JB, Rodan A, Kenyon C (1997) daf-16: An HNF-3/forkhead family member that can function to double the life-span of Caenorhabditis elegans. Science 278:1319–1322
Lindquist S, Craig EA (1988) The heat-shock proteins. Annu Rev Genet 22:631–677
Lu Q, Yang P, Huang X, Hu W, Guo B, Wu F, Lin L, Kovacs AL, Yu L, Zhang H (2011) The WD40 repeat PtdIns(3)P-binding protein EPG-6 regulates progression of omegasomes to autophagosomes. Dev Cell 21:343–357
Mair W, Dillin A (2008) Aging and survival: the genetics of life span extension by dietary restriction. Ann Rev Biochem 77:727–754
Mary J, Vougier S, Picot CR, Perichon M, Petropoulos I, Friguet B (2004) Enzymatic reactions involved in the repair of oxidized proteins. Exp Gerontol 39:1117–1123

McElwee JJ, Schuster E, Blanc E, Thomas JH, Gems D (2004) Shared transcriptional signature in Caenorhabditis elegans Dauer larvae and long-lived daf-2 mutants implicates detoxification system in longevity assurance. J Biol Chem 279:44533–44543
Menzies FM, Moreau K, Rubinsztein DC (2011) Protein misfolding disorders and macroautophagy. Curr Opin Cell Biol 23:190–197
Mizushima N, Yoshimori T, Levine B (2010) Methods in mammalian autophagy research. Cell 140:313–326
Morley JF, Morimoto RI (2004) Regulation of longevity in Caenorhabditis elegans by heat shock factor and molecular chaperones. Mol Biol Cell 15:657–664
Morris JZ, Tissenbaum HA, Ruvkun G (1996) A phosphatidylinositol-3-OH kinase family member regulating longevity and diapause in Caenorhabditis elegans. Nature 382:536–539
Mun JY, Lee TH, Kim JH, Yoo BH, Bahk YY, Koo HS, Han SS (2010) Caenorhabditis elegans mitofilin homologs control the morphology of mitochondrial cristae influence reproduction physiology. J Cell Physiol 224:748–756
Murphy CT, McCarroll SA, Bargmann CI, Fraser A, Kamath RS, Ahringer J, Li H, Kenyon C (2003) Genes that act downstream of DAF-16 to influence the lifespan of Caenorhabditis elegans. Nature 424:277–283
Nystrom T (2005) Role of oxidative carbonylation in protein quality control senescence. EMBO J 24:1311–1317
Ogg S, Paradis S, Gottlieb S, Patterson GI, Lee L, Tissenbaum HA, Ruvkun G (1997) The Forkheadtranscription factor DAF-16 transduces insulin-like metabolic longevitysignals in C elegans. Nature 389:995
Panowski SH, Wolff S, Aguilaniu H, Durieux J, Dillin A (2007) PHA-4/Foxa mediates diet-restriction-induced longevity of C elegans. Nature 447:550–555
Paradis S, Ruvkun G (1998) Caenorhabditis elegans Akt/PKB transduces insulin receptor-like signals from AGE-1 PI3 kinase to the DAF-16 transcription factor. Genes Dev 12:2488–2498
Paradis S, Ailion M, Toker A, Thomas JH, Ruvkun G (1999) A PDK1 homolog is necessary sufficient to transduce AGE-1 PI3 kinase signals that regulate diapause in Caenorhabditis elegans. Genes Dev 13:1438–1452
Park SK, Tedesco PM, Johnson TE (2009) Oxidative stress longevity in Caenorhabditis elegans as mediated by SKN-1. Aging Cell 8:258–269
Pickart CM (2001) Mechanisms underlying ubiquitination. Annual review of biochemistry 70:503–533
Pickart CM, Cohen RE (2004) Proteasomes their kin: proteases in the machine age. Nat Rev Mol Cell Biol 5:177–187
Pickford F, Masliah E, Britschgi M, Lucin K, Narasimhan R, Jaeger PA, Small S, Spencer B, Rockenstein E, Levine B et al (2008) The autophagy-related protein beclin 1 shows reduced expression in early Alzheimer disease regulates amyloid beta accumulation in mice. J Clin Invest 118:2190–2199
Ravikumar B, Sarkar S, Davies JE, Futter M, Garcia-Arencibia M, Green-Thompson ZW, Jimenez-Sanchez M, Korolchuk VI, Lichtenberg M, Luo S et al (2010) Regulation of mammalian autophagy in physiology pathophysiology. Physiol Rev 90:1383–1435
Rezzi S, Martin FP, Shanmuganayagam D, Colman RJ, Nicholson JK, Weindruch R (2009) Metabolic shifts due to long-term caloric restriction revealed in nonhuman primates. Exp Gerontol 44:356–362
Riederer BM, Leuba G, Vernay A, Riederer IM (2011) The role of the ubiquitin proteasome system in Alzheimer's disease. Exp Biol Med (Maywood) 236:268–276
Rubinsztein DC, Marino G, Kroemer G (2011) Autophagy aging. Cell 146:682–695
Scheffner M, Nuber U, Huibregtse JM (1995) Protein ubiquitination involving an E1-E2-E3 enzyme ubiquitin thioester cascade. Nature 373:81–83
Staub O, Rotin D (2006) Role of ubiquitylation in cellular membrane transport. Physiol Rev 86:669–707

Stefani M, Dobson CM (2003) Protein aggregation aggregate toxicity: new insights into protein folding, misfolding diseases and biological evolution. J Mol Med (Berl) 81:678–699
Steinkraus KA, Smith ED, Davis C, Carr D, Pendergrass WR, Sutphin GL, Kennedy BK, Kaeberlein M (2008) Dietary restriction suppresses proteotoxicity enhances longevity by an hsf-1-dependent mechanism in Caenorhabditis elegans. Aging Cell 7:394–404
Suzuki K, Ohsumi Y (2007) Molecular machinery of autophagosome formation in yeast Saccharomyces cerevisiae. FEBS Lett 581:2156–2161
Suzuki YJ, Carini M, Butterfield DA (2010) Protein carbonylation. Antioxid Redox Signal 12:323–325
Taylor RC, Dillin A (2011) Aging as an event of proteostasis collapse. Cold Spring Harb Perspect Biol 3. doi:10.1101/cshperspect.a004440
Tullet JMA, Hertweck M, An JH, Baker J, Hwang JY, Liu S, Oliveira RP, Baumeister R, Blackwell TK (2008) Direct inhibition of the longevity-promoting factor SKN-1 by insulin-like signaling in C. elegans. Cell 132:1025–1038
Tyedmers J, Mogk A, Bukau B (2010) Cellular strategies for controlling protein aggregation. Nat Rev Mol Cell Biol 11:777–788
Volovik Y, Maman M, Dubnikov T, Bejerano-Sagie M, Joyce D, Kapernick EA, Cohen E, Dillin A (2012) Temporal requirements of heat shock factor-1 for longevity assurance. Aging Cell 11:491–499
Walker GA, Lithgow GJ (2003) Lifespan extension in C. elegans by a molecular chaperone dependent upon insulin-like signals. Aging Cell 2:131–139
Weindruch R, Naylor PH, Goldstein AL, Walford RL (1988) Influences of aging and dietary restriction on serum thymosin alpha 1 levels in mice. J Gerontol 43:B40–42
Wolff S, Dillin A (2006) The trifecta of aging in Caenorhabditis elegans. Exp Gerontol 41:894–903
Wolff S, Ma H, Burch D, Maciel GA, Hunter T, Dillin A (2006) SMK-1, an essential regulator of DAF-16-mediated longevity. Cell 124:1039–1053
Wong E, Cuervo AM (2010) Integration of clearance mechanisms: the proteasome autophagy. Cold Spring Harb Perspect Biol 2:a006734
Yang Z, Klionsky DJ (2009) An overview of the molecular mechanism of autophagy. Curr Top Microbiol Immunol 335:1–32
Yokoyama K, Fukumoto K, Murakami T, Harada S, Hosono R, Wadhwa R, Mitsui Y, Ohkuma S (2002) Extended longevity of Caenorhabditis elegans by knocking in extra copies of hsp70F, a homolog of mot-2 (mortalin)/mthsp70/Grp75. FEBS Lett 516:53–57

The membrane sources of macroautophagy

Daniel J. Klionsky, Melinda A. Lynch-Day, Jiefei Geng, and Wei-Lien Yen

Abstract Macroautophagy (hereafter called autophagy) is a catabolic, homeostatic process that is ubiquitous in eukaryotes and is conserved from yeast to human. The morphological hallmark of autophagy is the formation of cytosolic double-membrane vesicles that sequester cytoplasm and deliver it to the vacuole (the analog of the lysosome), where the cargos are degraded and the breakdown products are released through membrane permeases for reuse. Autophagy is connected to a wide range of pathophysiological conditions. Accordingly, it has attracted considerable attention in the last decade. A major, still unresolved, question concerns the origin of the membrane(s) that contribute to the formation of the autophagosome and the mechanism through which this compartment is generated.

Introduction

The initial studies of autophagy focused on its role in response to the nutrient-related hormones insulin and glucagon (Deter et al. 1967; Deter and de Duve 1967); insulin, which signals the presence of glucose, elicits an inhibitory response, whereas glucagon is an inducer of autophagy. For several decades, the field of autophagy consisted primarily of morphological and pharmacological studies that relied heavily on electron microscopy and biochemistry (Klionsky 2007). Subsequently, the molecular analysis of autophagy began in yeast. Appropriately, these studies also focused on the role of this process in response to starvation.

The identification of autophagy-related genes in yeast led to their discovery in higher eukaryotes, which in turn ushered in an era of study marked by the

D.J. Klionsky (✉) • M.A. Lynch-Day • J. Geng • W.-L. Yen
Departments of Molecular, Cellular and Developmental Biology, Life Sciences Institute, University of Michigan, 210 Washtenaw Ave., Room 6036, Ann Arbor, MI, USA
e-mail: klionsky@umich.edu

R.I. Morimoto and Y. Christen (eds.), *Protein Quality Control in Neurodegenerative Diseases*, Research and Perspectives in Alzheimer's Disease,
DOI 10.1007/978-3-642-27928-7_4,

connection of autophagy to animal physiology and to various diseases (Huang and Klionsky 2007; Shintani and Klionsky 2004). For example, autophagy acts in tumor suppression, in the turnover of neuropeptides that can lead to certain types of neurodegeneration, and in the elimination of invasive microbes. Conversely, dysfunctional autophagy is associated with myopathies, liver disease and diabetes. Thus, there is tremendous interest in understanding the details of autophagy in order to manipulate the process for therapeutic purposes.

One of the unique aspects of autophagy is the mechanism of cargo sequestration. The autophagosome is generated de novo, in the sense that it does not bud off from a preexisting organelle as is the case with transient vesicles that deliver cargo throughout the secretory pathway. The process of autophagosome formation allows the vesicle to sequester cargo of essentially any size. Thus, autophagy is the primary mechanism for removing damaged or superfluous organelles.

Although much attention has focused on the autophagosome, it is a terminal structure that no longer has flexibility in terms of cargo capacity. It is actually the phagophore — the precursor to the autophagosome — that is the active membrane compartment (Klionsky 2011). An undefined nucleation event leads to the initial generation of the phagophore, which then expands through membrane addition to ultimately form the completed autophagosome. The outer membrane of the autophagosome then fuses with the limiting membrane of the vacuole, releasing the single-membrane interior vesicle into the lumen. The resulting autophagic body is broken down by vacuolar hydrolases to gain access to the cargo.

In addition to phagophore nucleation, there are many aspects of autophagosome biogenesis that are not understood. For example, two key issues concern the origin(s) of the membrane that form that autophagosome (Reggiori 2006) and the mechanism through which this membrane becomes part of the expanding phagophore. Recent studies, combining molecular techniques with electron microscopy, have started to address these questions.

The phagophore membrane is derived from multiple sources

Data regarding the source of the autophagic membrane for mammalian autophagy relied primarily on morphological electron microscopy-based studies. In brief, these studies suggested that membrane from multiple endomembrane compartments contribute to the autophagosome (Dunn 1990; Jaeken and Thines-Sempoux 1981; Locke and Sykes 1975). However, in the absence of molecular or genetic data to support these findings, the studies remained inconclusive. One fundamental difficulty was that the autophagosome membrane appeared to be almost devoid of membrane proteins (Fengsrud et al. 2000), which could otherwise provide information on their membrane of origin. This latter point has continued to complicate this question.

An undefined role for the secretory pathway in yeast autophagy was identified several years ago (Hamasaki et al. 2003; Ishihara et al. 2001; Reggiori et al. 2004b).

It was assumed that this reflected a general requirement for membrane flow, rather than that of a specific protein that transits through the secretory pathway. Furthermore, the endoplasmic reticulum (ER) is the main site of phospholipid synthesis, and it makes sense that this organelle would be involved in supplying membrane to meet the huge demand associated with autophagosome formation. Nonetheless, these studies did not provide information on the mechanism involved in membrane movement to the phagophore assembly site (PAS), the presumed location of phagophore nucleation and autophagosome formation.

Atg9 is a membrane marker

An important breakthrough came with the analysis of Atg9. When tagged with a fluorophore such as green fluorescent protein (GFP), most Atg proteins in yeast are detected in cytosolic pools and as a punctum that corresponds to the PAS. In contrast, Atg9-GFP exists in multiple puncta, only one of which corresponds to the PAS (Noda et al. 2000). Atg9 is one of the few transmembrane Atg proteins, and the only one that is absolutely required for autophagy. To determine the nature of the peripheral Atg9 puncta, localization of Atg9-GFP was examined in strains defective in each of the individual *ATG* genes. One important observation from this analysis was the finding that, in strains deleted for *ATG1*, Atg9-GFP localized exclusively at the PAS (Reggiori et al. 2004a). Extending these studies with a conditional, temperature sensitive *atg1* mutant suggests that Atg9 moves between these peripheral pools and the PAS. At a nonpermissive temperature, Atg9-GFP is detected only at the PAS, but soon after shifting to a permissive temperature it is detected in multiple puncta, although cycling has not been directly demonstrated by time-lapse imaging. Coupled with its nature as a membrane protein, these data led to the hypothesis that Atg9 is a membrane carrier for phagophore expansion. Thus, Atg9 serves as a potential marker for the membrane that forms the autophagosome.

In addition to Atg1, Atg13, which is required for maximal Atg1 kinase activity (Kamada et al. 2000), Atg2 and Atg18 (Guan et al. 2001; Wang et al. 2001), which form a complex, and Atg14 (Kametaka et al. 1998), which is a component of the PtdIns 3-kinase complex (Kihara et al. 2001), are also required for the retrograde transport of Atg9 from the PAS back to the peripheral sites (Reggiori et al. 2004a). The function of Atg2 and Atg18 is not known; however, Atg18 is a PtdIns(3)P binding protein (Guan et al. 2001), and the role of Atg14 with regard to Atg9 movement may be due to the need to recruit Atg18 to the PAS.

Two other proteins, Atg23 (Tucker et al. 2003) and Atg27 (Yen et al. 2007), also display a multiple puncta distribution. Subsequent studies demonstrated that these proteins travel to and from the PAS along with Atg9 (Legakis et al. 2007). The requirement for these components in anterograde transport to the PAS was determined through the use of an epistasis assay that relies on the phenotype of the *atg1Δ* strain. The Transport of Atg9 after Knocking out *ATG1*, or TAKA assay, combines a second mutation with *atg1Δ* and examines the effect of the second mutation on

Atg9 localization (Cheong et al. 2005). If the double mutant displays a multiple puncta phenotype, the protein encoded by the second gene likely acts prior to Atg1 with regard to Atg9 movement. Thus, *atg23Δ* and *atg27Δ* mutations, when combined with *atg1Δ,* result in cells that have multiple Atg9-GFP puncta.

One additional protein involved in anterograde movement of Atg9 is Atg11 (He et al. 2006; Kim et al. 2001). The Atg11 protein may act as a scaffold that interacts with several other autophagy proteins and also self-interacts (Yorimitsu and Klionsky 2005). One possibility is that Atg11 tethers Atg9-containing membrane to the PAS due to Atg11 self-interactions, although Atg9 self-interaction is also critical for efficient movement to the PAS (He et al. 2008). Atg9 forms clusters that may play an important role in its function, although further studies are needed to determine the nature of the Atg9-containing complex. Finally, the actin cytoskeleton and components of the Arp2/3 complex are needed for proper movement of Atg9 (Monastyrska et al. 2008).

Initial studies of Atg9-GFP indicated that the protein was located near the mitochondrial reticulum (Reggiori et al. 2005). Recent immunoelectron microscopy studies reveal that Atg9 is located in tubulovesicular clusters that are near, but not connected with, the mitochondria (Mari et al. 2010). These Atg9 reservoirs, which correspond to the peripheral population of Atg9, may generate the PAS and are likely precursors to the phagophore (Fig. 1).

Trafficking Machinery in the Secretory Pathway Is Utilized for Autophagy

The Atg9 reservoirs appear to be one of the sources of membrane used for phagophore formation/expansion; however, it is not known how Atg9 is delivered to these sites. As a membrane protein, Atg9 transits through the ER and Golgi complex.

Movement of proteins and membranes through the early secretory pathway is regulated by Rab GTPases. These proteins act as molecular switches, aiding in multiple steps of membrane traffic, including membrane fission, vesicle tethering, and membrane fusion. Rab GTPases cycle between an inactive state that is bound to GDP and an active state that is bound to GTP. Cycling to the active state is aided by a guanine nucleotide exchange factor (GEF). In yeast, ER-to-Golgi and intra-Golgi trafficking is regulated by the Rab GTPase Ypt1 and its GEF, TRAPPI. There are two main TRAPP complexes involved in the secretory pathway. TRAPPI is a multimeric complex comprised of six subunits (Bet3, Bet5, Trs20, Trs23, Trs31 and Trs33). TRAPPII is responsible for regulating Golgi-to-endosome trafficking and contains three additional subunits (Trs65, Trs120, and Trs130). Previously, Trs85 was considered to be a basic subunit and part of both TRAPP complexes. A screen of the yeast knockout library identified the *trs85Δ* strain as being defective in autophagy and the cytoplasm to vacuole targeting (Cvt) pathway (Nazarko et al.

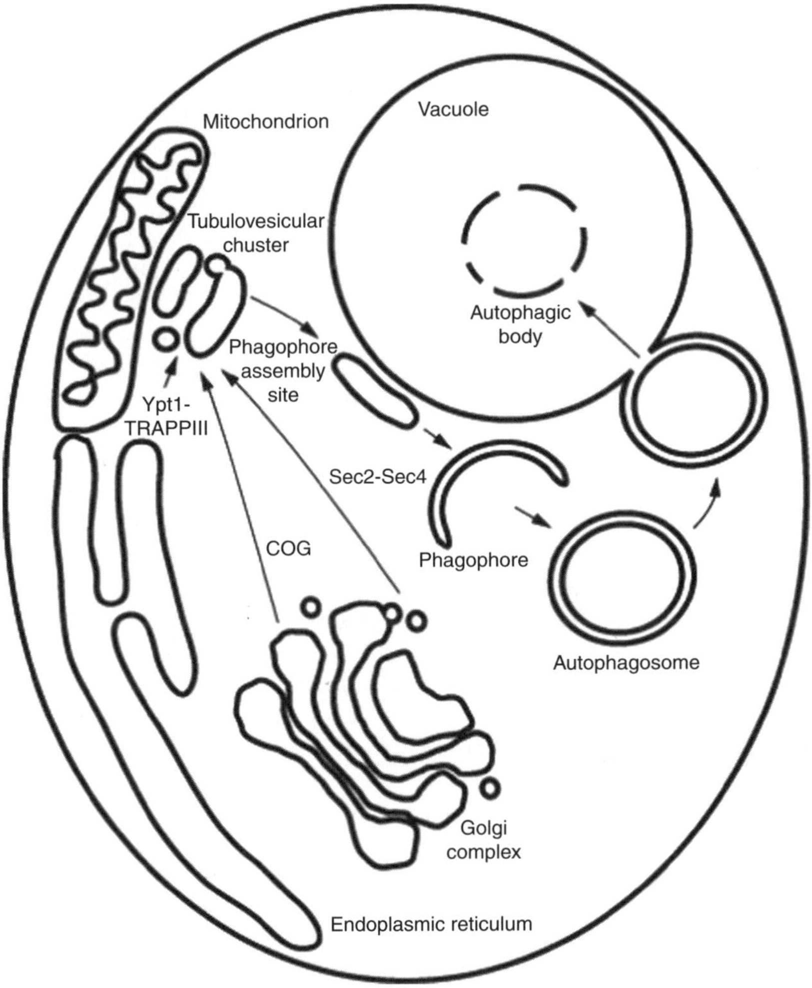

Fig. 1 Schematic model for autophagosome formation in yeast. Multiple compartments provide membrane that is used to generate the autophagosome. These include the endoplasmic reticulum (via the Rab protein Ypt1 and its GEF, TRAPPIII), the Golgi complex (involving the COG complex as a tether), and the trans-Golgi network (utilizing the Rab protein Sec4 and its GEF, Sec2). These organelles of the secretory pathway may direct membrane that contains Atg9 to the tubulovesicular clusters (Atg9 reservoirs) that are located near mitochondria. The Atg9 reservoirs in turn may be critical in biogenesis of the perivacuolar phagophore assembly site that serves as the location for nucleation of the initial sequestering compartment, the phagophore, which is the precursor to the autophagosome. After completion, the autophagosome fuses with the vacuole, releasing the autophagic body into the lumen. Breakdown of the autophagic body and release of the degraded cargo into the cytosol constitute the last steps of macroautophagy

2005). Although *TRS33*, *TRS65* and *TRS85* are all nonsessential genes, *TRS85* was the only gene encoding a nonessential TRAPP subunit that was identified in the screen, and subsequent studies verified that Trs33 and Trs65 are not required for autophagy. These findings raised the possibility that Trs85 functions independently of TRAPP or that it is part of a unique TRAPP complex. Subsequent studies revealed the presence of a third complex, TRAPPIII, which is the only TRAPP complex that contains Trs85 (instead of Trs65, Trs120 and Trs130) and is specific to autophagy (Lynch-Day et al. 2010). The *trs85Δ* mutant has no effect on ER-to-Golgi transport in *Saccharomyces cerevisiae*. Thus, Trs85 diverts this complex into a role in autophagy rather than secretory pathway transport (Fig. 1).

The TRAPP complex functions as a GEF for the Rab protein Ypt1, which is also required for autophagy (Lynch-Day et al. 2010). The defect in autophagy seen in *ypt1* mutants is not an indirect result of blocking the early secretory pathway because the phenotype is still observed when secretory traffic is normal. Both Trs85 and Ypt1 localize to the PAS, and Ypt1 colocalizes in part with Atg9. Accordingly, one hypothesis is that Ypt1-TRAPPIII is involved in delivering membrane directly from the ER to the PAS.

Similar to the adoption of the Ypt1-TRAPP complex for autophagy, other trafficking machinery is utilized for this process under starvation conditions. Another example is seen with the conserved oligomeric Golgi (COG) complex that functions as a tether and is mainly involved in transport within the Golgi complex (Yen et al. 2010). The yeast COG complex is composed of two lobes: lobe A (formed of the Cog2, Cog3 and Cog4 subunits) and lobe B (comprised of Cog5, Cog6, Cog7 and Cog8) with Cog1 at the center (Kim et al. 1999; Ram et al. 2002; Suvorova et al. 2002; VanRheenen et al. 1999; Whyte and Munro 2001). Lobe A, but not lobe B, subunits of the COG complex are essential and they are required for autophagy, whereas all the subunits are required for the Cvt pathway (Yen et al. 2010).

The *cog* mutants defective in autophagy display aberrant localization of Atg9, having a partial block in anterograde movement. In addition, there is a defect in the localization of Atg8, which forms small clusters but is dispersed throughout the cytoplasm. The latter finding suggests that the COG complex might function as a tether for Cvt vesicle and autophagosome formation and completion. Furthermore, COG complex subunits localize to the PAS and interact with Atg proteins, including Atg9 (Yen et al. 2010). These data further implicate the COG complex in tethering Atg9-containing vesicles at the PAS and/or tubulovesicular cluster (the Atg9 reservoirs). Together, these observations indicate that the Golgi complex may contribute to autophagosome formation (Fig. 1).

Defects in ER-to-Golgi and intra-Golgi trafficking potentially have indirect effects on autophagy because of the requirement for the secretory pathway in this process. In contrast, there is no indication that secretion itself is needed for autophagy. Nonetheless, mutations in the *SEC2* or *SEC4* gene cause defects in autophagy (Geng et al. 2010). The Rab protein Sec4 and its GEF Sec2 are involved in the movement of secretory vesicles from the *trans*-Golgi network (TGN) to the plasma membrane. Yeast strains with mutations in these genes display defects in Atg9 localization, suggesting that the TGN is another source of Atg9-containing membrane. In addition, mutations in Ypt31 and Ypt32 result in a block of autophagy (Geng et al. 2010). Ypt31 and Ypt32 are two redundant Rab proteins and they are essential for the exit of secretory vesicles from the TGN. Therefore, the autophagic defect in a *ypt31Δ ypt32*^{ts} mutant also supports the hypothesis that the TGN is one of the membrane sources for autophagosome formation. Coupled with the fact that protein secretion is substantially attenuated during starvation (Geng et al. 2010; Shorer et al. 2005), these data suggest that Sec2/4-mediated membrane flow is shared between the secretory pathway and autophagy. Transport may be converted from one pathway to another when the nutrient level changes. In nutrient-rich conditions, Sec2 and Sec4 mediate secretory vesicle delivery to the plasma membrane to support cell growth; when the cells encounter a shortage of nutrients, cell growth is halted while this membrane flow is directed to autophagosome formation (Fig. 1). Although the nutrient-dependent regulatory mechanism remains unknown, these observations indicate that lipid movement among the ER, Golgi complex and TGN contributes to expansion of the phagophore.

Conclusions

The observations discussed above indicate that multiple compartments likely supply membrane for autophagosome formation in yeast. Returning briefly to mammalian cells, data from various laboratories have implicated the mitochondria (Hailey et al. 2010), the TGN/endosome (Young et al. 2006), the ER (Axe et al. 2008) and the plasma membrane (Ravikumar et al. 2010) as sources of the autophagosome membrane. These latter studies are based on fluorescence microscopy and molecular genetic techniques. Thus, we have come full circle, starting with the initial electron microscopy studies that suggested that multiple membranes serve as sources for the autophagosome.

References

Axe EL, Walker SA, Manifava M, Chandra P, Roderick HL, Habermann A, Griffiths G, Ktistakis NT (2008) Autophagosome formation from membrane compartments enriched in phosphatidylinositol 3-phosphate and dynamically connected to the endoplasmic reticulum. J Cell Biol 182:685–701

Cheong H, Yorimitsu T, Reggiori F, Legakis JE, Wang C-W, Klionsky DJ (2005) Atg17 regulates the magnitude of the autophagic response. Mol Biol Cell 16:3438–3453

Deter RL, de Duve C (1967) Influence of glucagon, an inducer of cellular autophagy, on some physical properties of rat liver lysosomes. J Cell Biol 33:437–449

Deter RL, Baudhuin P, de Duve C (1967) Participation of lysosomes in cellular autophagy induced in rat liver by glucagon. J Cell Biol 35:C11–16

Dunn WA Jr (1990) Studies on the mechanisms of autophagy: formation of the autophagic vacuole. J Cell Biol 110:1923–1933

Fengsrud M, Erichsen ES, Berg TO, Raiborg C, Seglen PO (2000) Ultrastructural characterization of the delimiting membranes of isolated autophagosomes and amphisomes by freeze-fracture electron microscopy. Eur J Cell Biol 79:871–882

Geng J, Nair U, Yasumura-Yorimitsu K, Klionsky DJ (2010) Post-Golgi Sec proteins are required for autophagy in *Saccharomyces cerevisiae*. Mol Biol Cell 21:2257–2269

Guan J, Stromhaug PE, George MD, Habibzadegah-Tari P, Bevan A, Dunn WA Jr, Klionsky DJ (2001) Cvt18/Gsa12 is required for cytoplasm-to-vacuole transport, pexophagy, and autophagy in *Saccharomyces cerevisiae* and *Pichia pastoris*. Mol Biol Cell 12:3821–3838

Hailey DW, Rambold AS, Satpute-Krishnan P, Mitra K, Sougrat R, Kim PK, Lippincott-Schwartz J (2010) Mitochondria supply membranes for autophagosome biogenesis during starvation. Cell 141:656–667

Hamasaki M, Noda T, Ohsumi Y (2003) The early secretory pathway contributes to autophagy in yeast. Cell Struct Funct 28:49–54

He C, Song H, Yorimitsu T, Monastyrska I, Yen W-L, Legakis JE, Klionsky DJ (2006) Recruitment of Atg9 to the preautophagosomal structure by Atg11 is essential for selective autophagy in budding yeast. J Cell Biol 175:925–935

He C, Baba M, Cao Y, Klionsky DJ (2008) Self-interaction is critical for Atg9 transport and function at the phagophore assembly site during autophagy. Mol Biol Cell 19:5506–5516

Huang J, Klionsky DJ (2007) Autophagy and human disease. Cell Cycle 6:1837–1849

Ishihara N, Hamasaki M, Yokota S, Suzuki K, Kamada Y, Kihara A, Yoshimori T, Noda T, Ohsumi Y (2001) Autophagosome requires specific early Sec proteins for its formation and NSF/SNARE for vacuolar fusion. Mol Biol Cell 12:3690–3702

Jaeken L, Thines-Sempoux D (1981) A three-dimensional study of organelle interrelationships in regenerating rat liver. 5. Autophagy. Cell Biol Int Rep 5:253–260

Kamada Y, Funakoshi T, Shintani T, Nagano K, Ohsumi M, Ohsumi Y (2000) Tor-mediated induction of autophagy via an Apg1 protein kinase complex. J Cell Biol 150:1507–1513

Kametaka S, Okano T, Ohsumi M, Ohsumi Y (1998) Apg14p and Apg6/Vps30p form a protein complex essential for autophagy in the yeast, *Saccharomyces cerevisiae*. J Biol Chem 273:22284–22291

Kihara A, Noda T, Ishihara N, Ohsumi Y (2001) Two distinct Vps34 phosphatidylinositol 3-kinase complexes function in autophagy and carboxypeptidase Y sorting in *Saccharomyces cerevisiae*. J Cell Biol 152:519–530

Kim DW, Sacher M, Scarpa A, Quinn AM, Ferro-Novick S (1999) High-copy suppressor analysis reveals a physical interaction between Sec34p and Sec35p, a protein implicated in vesicle docking. Mol Biol Cell 10:3317–3329

Kim J, Kamada Y, Stromhaug PE, Guan J, Hefner-Gravink A, Baba M, Scott SV, Ohsumi Y, Dunn WA Jr, Klionsky DJ (2001) Cvt9/Gsa9 functions in sequestering selective cytosolic cargo destined for the vacuole. J Cell Biol 153:381–396

Klionsky DJ (2007) Autophagy: from phenomenology to molecular understanding in less than a decade. Nat Rev Mol Cell Biol 8:931–937

Klionsky DJ (2011) The autophagosome is overrated! Autophagy 7:353–354

Legakis JE, Yen W-L, Klionsky DJ (2007) A cycling protein complex required for selective autophagy. Autophagy 3:422–432

Locke M, Sykes AK (1975) The role of the Golgi complex in the isolation and digestion of organelles. Tissue Cell 7:143–158

Lynch-Day MA, Bhandari D, Menon S, Huang J, Cai H, Bartholomew CR, Brumell JH, Ferro-Novick S, Klionsky DJ (2010) Trs85 directs a Ypt1 GEF, TRAPPIII, to the phagophore to promote autophagy. Proc Natl Acad Sci USA 107:7811–7816

Mari M, Griffith J, Rieter E, Krishnappa L, Klionsky DJ, Reggiori F (2010) An Atg9-containing compartment that functions in the early steps of autophagosome biogenesis. J Cell Biol 190:1005–1022

Monastyrska I, He C, Geng J, Hoppe AD, Li Z, Klionsky DJ (2008) Arp2 links autophagic machinery with the actin cytoskeleton. Mol Biol Cell 19:1962–1975

Nazarko TY, Huang J, Nicaud JM, Klionsky DJ, Sibirny AA (2005) Trs85 is required for macroautophagy, pexophagy and cytoplasm to vacuole targeting in *Yarrowia lipolytica* and *Saccharomyces cerevisiae*. Autophagy 1:37–45

Noda T, Kim J, Huang W-P, Baba M, Tokunaga C, Ohsumi Y, Klionsky DJ (2000) Apg9p/Cvt7p is an integral membrane protein required for transport vesicle formation in the Cvt and autophagy pathways. J Cell Biol 148:465–480

Ram RJ, Li B, Kaiser CA (2002) Identification of Sec36p, Sec37p, and Sec38p: components of yeast complex that contains Sec34p and Sec35p. Mol Biol Cell 13:1484–1500

Ravikumar B, Moreau K, Jahreiss L, Puri C, Rubinsztein DC (2010) Plasma membrane contributes to the formation of pre-autophagosomal structures. Nat Cell Biol 12:747–757

Reggiori F (2006) Membrane origin for autophagy. Curr Top Dev Biol 74:1–30

Reggiori F, Tucker KA, Stromhaug PE, Klionsky DJ (2004a) The Atg1-Atg13 complex regulates Atg9 and Atg23 retrieval transport from the pre-autophagosomal structure. Dev Cell 6:79–90

Reggiori F, Wang C-W, Nair U, Shintani T, Abeliovich H, Klionsky DJ (2004b) Early stages of the secretory pathway, but not endosomes, are required for Cvt vesicle and autophagosome assembly in *Saccharomyces cerevisiae*. Mol Biol Cell 15:2189–2204

Reggiori F, Shintani T, Nair U, Klionsky DJ (2005) Atg9 cycles between mitochondria and the pre-autophagosomal structure in yeasts. Autophagy 1:101–109

Shintani T, Klionsky DJ (2004) Autophagy in health and disease: a double-edged sword. Science 306:990–995

Shorer H, Amar N, Meerson A, Elazar Z (2005) Modulation of N-ethylmaleimide-sensitive factor activity upon amino acid deprivation. J Biol Chem 280:16219–16226

Suvorova ES, Duden R, Lupashin VV (2002) The Sec34/Sec35p complex, a Ypt1p effector required for retrograde intra-Golgi trafficking, interacts with Golgi SNAREs and COPI vesicle coat proteins. J Cell Biol 157:631–643

Tucker KA, Reggiori F, Dunn WA Jr, Klionsky DJ (2003) Atg23 is essential for the cytoplasm to vacuole targeting pathway and efficient autophagy but not pexophagy. J Biol Chem 278:48445–48452

VanRheenen SM, Cao X, Sapperstein SK, Chiang EC, Lupashin VV, Barlowe C, Waters MG (1999) Sec34p, a protein required for vesicle tethering to the yeast Golgi apparatus, is in a complex with Sec35p. J Cell Biol 147:729–742

Wang C-W, Kim J, Huang W-P, Abeliovich H, Stromhaug PE, Dunn WA Jr, Klionsky DJ (2001) Apg2 is a novel protein required for the cytoplasm to vacuole targeting, autophagy, and pexophagy pathways. J Biol Chem 276:30442–30451

Whyte JR, Munro S (2001) The Sec34/35 Golgi transport complex is related to the exocyst, defining a family of complexes involved in multiple steps of membrane traffic. Dev Cell 1:527–537

Yen W-L, Legakis JE, Nair U, Klionsky DJ (2007) Atg27 is required for autophagy-dependent cycling of Atg9. Mol Biol Cell 18:581–593

Yen W-L, Shintani T, Nair U, Cao Y, Richardson BC, Li Z, Hughson FM, Baba M, Klionsky DJ (2010) The conserved oligomeric Golgi complex is involved in double-membrane vesicle formation during autophagy. J Cell Biol 188:101–114

Yorimitsu T, Klionsky DJ (2005) Atg11 links cargo to the vesicle-forming machinery in the cytoplasm to vacuole targeting pathway. Mol Biol Cell 16:1593–1605

Young ARJ, Chan EYW, Hu XW, Köchl R, Crawshaw SG, High S, Hailey DW, Lippincott-Schwartz J, Tooze SA (2006) Starvation and ULK1-dependent cycling of mammalian Atg9 between the TGN and endosomes. J Cell Sci 119:3888–3900

Selective autophagy in cellular quality control

Susmita Kaushik and Ana Maria Cuervo

Abstract Cells count on surveillance systems to handle protein alterations and organelle damage. This chapter focuses on one of the cellular systems for protein removal, autophagy, or the degradation of intracellular components in lysosomes, and its relationship with the toxicity induced by aberrant proteins. We describe some of the recent findings on the interplay between pathogenic proteins and autophagy, looking at it as a two-sided relationship. In the first part, we comment on the contribution of the different autophagic pathways to the removal of pathogenic proteins; in the second part, we review the toxic effect that pathogenic proteins can exert on the activity of different autophagic pathways.

Introduction

As described in other chapters of this book, autophagy is a conserved catabolic pathway by which cells deliver intracellular components – proteins and organelles – to lysosomes for degradation (Yang and Klionsky 2010a). This continuous turnover is required to maintain cellular homeostasis and, along with the chaperone network, autophagy contributes to cellular quality control (Morimoto and Cuervo 2009). Functions of the autophagic pathway extend beyond cellular cleaning and maintenance. Autophagy has been shown to participate in cellular differentiation, tissue remodeling as a pro-survival mechanism, and in the defense against pathogens (Mizushima et al. 2008). In this chapter, we will focus preferentially on the function of selective forms of autophagy in cellular quality control, with emphasis on the interplay between proteotoxicity and the autophagic system.

S. Kaushik • A.M. Cuervo (✉)
Department of Developmental and Molecular Biology, Department of Medicine and Institute for Aging Studies, Albert Einstein College of Medicine, Bronx, NY 10461, USA
e-mail: ana–maria.cuervo@einstein.yu.edu

R.I. Morimoto and Y. Christen (eds.), *Protein Quality Control in Neurodegenerative Diseases*, Research and Perspectives in Alzheimer's Disease,
DOI 10.1007/978-3-642-27928-7_5,

The term proteotoxicity is commonly used to reflect the harmful effect in cells of different types of misbehaving proteins, including proteins that are misfolded, mutated or modified in any way that favors non-specific interactions with other proteins (Morimoto 2008). Chaperones and the proteolytic systems constitute the main mechanisms of defense against these proteins, which otherwise would accumulate inside cells and contribute in the long run to cellular malfunction and often to cellular death. Numerous human disorders, generically known as protein conformational diseases, originate as a result of the accumulation of pathogenic proteins in different organs and tissues (Balch et al. 2008). This type of disease encompasses, among others, neurodegenerative conditions such as Alzheimer's disease (AD) or Parkinson's disease (PD). However, the toxic effect of pathogenic proteins is not restricted to the central nervous system. In general, accumulation of toxic proteins in any type of differentiated, non-dividing cell could cause cellular injury if it is not efficiently handled by the quality control mechanisms.

In this chapter, we review the relationship of the autophagic system with protein conformational disorders and pathogenic proteins from opposite angles. On one hand, we will address the capacity of the autophagic system to become an efficient mechanism for clearance of pathogenic proteins and, on the other hand, we will analyze autophagy as a target of the toxicity of pathogenic proteins.

Autophagic pathways in mammalian cells

Cells can perform autophagy through different mechanisms that basically differ in the way in which cytosolic cargo is delivered to lysosomes for degradation. In macroautophagy, the better characterized type of autophagy, cytosolic components are sequestered by a de novo formed membrane that seals to form a double membrane vesicle or autophagosome (Yang and Klionsky 2009, 2010b). This vesicle is then infused with the hydrolases required for cargo degradation upon fusion with endosomes and lysosomes. A similar process of sequestration of cytosolic regions takes place during microautophagy, but in this case it is the lysosomal membrane itself that invaginates to internalize the cytosolic components inside single membrane vesicles that, once pinched off from the membrane, are rapidly degraded in the lysosomal lumen (Marzella et al. 1981). Direct traffic of cytosolic proteins across the lysosomal membrane for degradation also occurs in cells through an autophagic variant known as chaperone-mediated autophagy (Arias and Cuervo 2011; Fig. 1).

The molecular basis of macroautophagy has been extensively reviewed in other chapters in this book. Briefly, a subset of about 32 genes, known as autophagy-related genes (or ATG), code for proteins that participate in each of the steps of macroautophagy (Yang and Klionsky 2009, 2010b). These Atg proteins organize in multimeric complexes to mediate the nucleation of the membrane that surrounds cargo, through the shuttling of lipids and proteins from other intracellular membranes to the site of its formation. The coordinated action of other protein

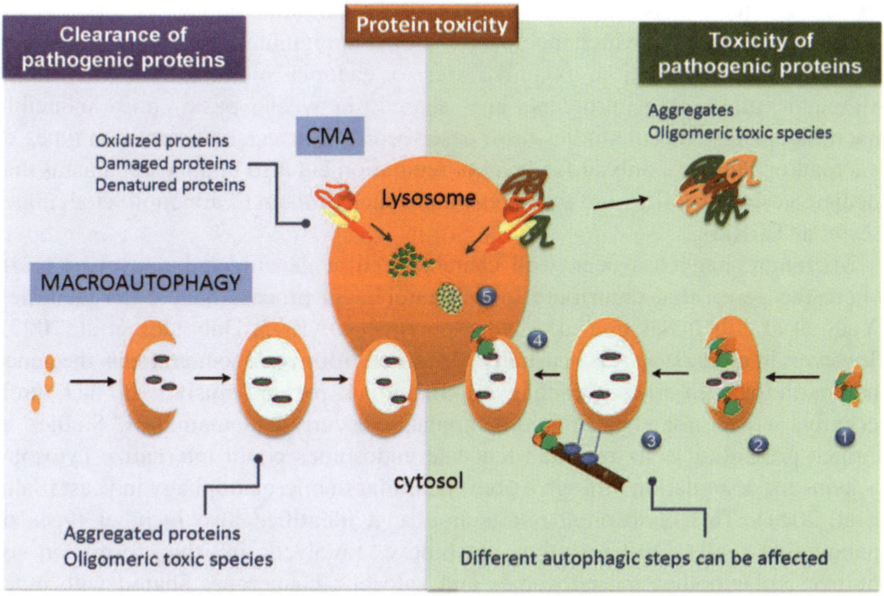

Fig. 1 The two-sided relationship between autophagic pathways and proteotoxicity. **Left:** Macroautophagy and chaperone-mediated autophagy (CMA) participate actively in the removal of different types of toxic proteins. While CMA can remove the altered proteins in a very selective manner, a main limitation of this process is that only soluble proteins amenable to complete unfolding can be subjected to degradation through this pathway. Once proteins are organized into irreversible oligomeric complexes or in aggregates, macroautophagy is the only pathway able to clear up these protein products upon their sequestration from the cytosol. **Right:** Macroautophagy and chaperone-mediated autophagy are both targets of toxicity mediated by proteins. Abnormal interactions of pathogenic proteins with components of the CMA translocation machinery on the surface of the lysosome prevent their degradation through this pathway and at the same time block degradation of other proteins by CMA. Pathogenic proteins can alter normal macroautophagy functioning at multiple steps of this process: *1* initiation and formation of the autophagosome, *2* cargo recognition, *3* vesicular trafficking of the autophagosomes toward the lysosomes, *4* fusion of the autophagic and lysosomal compartments and *5* degradation of cargo. Examples of each of these types of blockage are provided in the text

complexes guarantees the elongation of the membrane decorated on both sides by the light chain protein type 3 (LC3), a protein that remains bound to the autophagosome and that has become the most utilized marker for this compartment (Asanuma et al. 2003). Autophagosomes move through microtubules to reach the endosomal and lysosomal compartments that are carriers of the degradative enzymes. Different types of proteins participate in the fusion of autophagosomes with other vesicles, but the precise mechanisms of fusion remain, for the most part, unknown (Yang and Klionsky 2009, 2010b). Different forms of macroautophagy have been described, depending on the cargo sequestered (mitophagy, pexophagy, aggregophagy) or the mechanisms that determine the initiation of the autophagic process (basal or inducible). Basal macroautophagy is often considered responsible for cellular quality control whereas the term inducible autophagy is reserved for those circumstances in which macroautophagy is upregulated in response to cellular stressors such as starvation, oxidative stress or endoplasmic reticulum (ER) stress. Although most Atg proteins are shared between basic and inducible macroautophagy, recent studies have described differences between both types of macroautophagy, not only at the level of regulation but also in the mechanisms that mediate vesicular traffic and fusion events in each of them (Yamamoto et al. 2006; Lee et al. 2010b).

Microautophagy has been well characterized at the molecular level in yeast, where the genes that contribute to this autophagic process have been identified (Yuan et al. 1997; Sakai et al. 1998; Roberts et al. 2003; Dubouloz et al. 2005). However, in contrast to ATGs usually conserved from yeast to mammals, the genes that participate in microautophagy in mammals remain elusive. In fact, until recently, it was not clear if microautophagy occurred in mammals. Studies in antigen-presenting cells revealed that late endosomes could internalize cytosolic proteins for degradation through a process similar to microautophagy in yeast (Sahu et al. 2011). This endosomal microautophagy, identified also in other types of mammalian cells, utilizes the machinery involved in the formation of multivesicular bodies in endosomes and cytosolic chaperones shared with other autophagic processes (Sahu et al. 2011).

Chaperone-mediated autophagy (CMA) contributes to the degradation of a subset of single soluble cytosolic proteins that bear a targeting motif in their amino acid sequence (Dice 2007; Cuervo 2010). A cytosolic chaperone, hsc70, the constitutive member of the hsp70 family of molecular chaperones, binds to this motif and brings the substrate protein to the surface of the lysosomes (Chiang et al. 1989). Once there, the protein/substrate complex binds to the lysosome-associated membrane protein type 2A (LAMP-2A), which has a dual function in binding and translocation of substrates into lysosomes (Bandyopadhyay et al. 2008, 2010). Binding of substrates to LAMP-2A drives their organization into a multimeric complex required for substrate translocation. A lysosomal variant of hsc70 resident in the lysosomal lumen is also required to complete the translocation step, which is followed by rapid degradation of the substrate protein in the lysosomal lumen. Certain levels of basal CMA activity can be detected in almost all type of cells, but this pathway is maximally activated under conditions of stress, such as prolonged nutritional deprivation (beyond 10 h), oxidative stress, toxic exposure, and stress conditions resulting in preferential damage of soluble proteins with minimal organelle compromise, as organelles cannot be degraded through CMA (Dice 2007; Cuervo 2010).

The lack of tools for the study of microautophagy in mammals has prevented the establishment of clear connections of this autophagic process and proteotoxicity. In contrast, both macroautophagy and CMA have been shown to contribute to the cellular defense against protein toxicity and are also common targets of pathogenic proteins in different protein conformational disorders. This reciprocal interaction is described in detail in the following sections.

Autophagy in the fight against proteotoxicity

Early reports regarding the ability of macroautophagy to degrade proteins when present as irreversible oligomeric complexes or as aggregates inside cells generated a considerable level of excitement in fields such as neurodegeneration (Ravikumar et al. 2002). Macroautophagy revealed itself as an efficient mechanism for the elimination of protein structures no longer amenable for degradation through the ubiquitin-proteasome system and, in fact, chemical upregulation of macroautophagy proved beneficial in flies and mouse models of specific protein conformational disorders (Sarkar et al. 2007). Since these early reports, the list of pathological conditions in which macroautophagy has been described to be upregulated and contributes to the removal of toxic protein adducts keeps growing, and different manipulations aimed at upregulating macroautophagy for aggregate removal have succeeded in slowing down the course of some of these disorders, at least in their corresponding animal models (Sarkar et al. 2007, 2009). These findings have generated a marked interest in elucidating the mechanism(s) by which these structures are selectively targeted for macroautophagic degradation, now known as aggregophagy.

Components of macroautophagy are often detected in close association to the aggregated proteins, and they were initially thought to be a mere manifestation of their non-specific entrapment inside these protein complexes along with many other cytosolic proteins (Ravikumar et al. 2004; Iwata et al. 2005). However, a closer analysis of the Atgs in aggregates revealed that they are indeed a manifestation of the formation of autophagosomes around these inert structures. The mechanisms that contribute to the selective recognition of aggregates by macroautophagy are currently the subject of intense investigation, as it is perceived that, rather than global upregulation of macroautophagy, which could result in non-selective unwanted degradation of necessary cellular components, manipulations aimed at selectively enhancing aggregophagy could be beneficial in the clinical setting (Kirkin et al. 2009; Novak and Dikic 2011). As is the case for the selective targeting of organelles by macroautophagy, specific cargo-recognition molecules have been shown to act as links between the aggregates and the components of the macroautophagic machinery in order to generate a "customized" limited membrane around the aggregates. Two proteins, p62 and NBR1 (neighbor of BRCA1 gene 1), are among the best characterized cargo-recognition molecules for aggregophagy (Lamark et al. 2009). These proteins contain both an LC3-interacting region (LIR) and a ubiquitin-binding region that acts as a cargo-recognition platform (Noda et al. 2010). In fact, ubiquitin, the protein moiety that when conjugated to soluble proteins targets them for degradation through the ubiquitin-proteasome system, has also been shown to contribute to tagging of protein aggregates for degradation, in this case, via macroautophagy. The distinction between proteasomal and lysosomal targeting may be dictated by the type of linkage used by ubiquitin to conjugate to the substrate protein (Dikic et al. 2010). Conjugation of ubiquitin at its lysine 48 (K48) is the most common tag for proteasomal degradation whereas

K63 linkage or a combination of K48/K63 linkages – likely as mixed chains on the surface of the aggregates – seems the preferred link for p62 and NBR1. The series of events that occur from the binding of the cargo-recognition molecules to the surface of the protein aggregates until the formation of the surrounding limiting membrane are currently being scrutinized. Recent studies indicate that the presence of p62 and NBR1 in aggregates is necessary for their macroautophagic removal but it is not sufficient. Although a growing list of pathogenic proteins has been shown to undergo macroautophagic degradation once organized into aggregates, autophagic removal of aggregates may not follow the mere binding of cargo-recognition proteins (Wong et al. 2008). Thus, in the same experimental setting in which aggregates of α-synuclein, the main component of Lewy bodies in PD neurons, or of mutant huntingtin can be easily removed by macroautophagy, aggregates of proteins such as p38, which also accumulate in PD inclusions or desmin and are present in the aggregates of muscle dystrophies, remain undegraded even when macroautophagy is maximally activated. Failure to eliminate these aggregates by macroautophagy is not a result of poor degradation through this pathway; rather, these aggregates seem to completely escape the surveillance of the autophagic system as they cannot even be detected in autophagosomes (Wong et al. 2008). In support of the hypothesis that cargo-recognition molecules are necessary but not sufficient, aggregates of p38 and desmin are positive for the presence of p62. It is possible that the way in which p62 associates with the pathogenic proteins in one type of aggregates versus the other is different and that the dynamics of association/dissociation of this molecule from the surface of protein inclusions, as previously shown for chaperones (Prahlad et al. 2008), is an important determinant of its ability to bring along the machinery of the autophagic system. In that respect, current efforts are focused on elucidating the order of "arrival" of the different protein complexes required for the formation of the autophagosome to the protein aggregates.

Macroautophagy may also contribute to reducing the intracellular content of aggregated proteins, not only through their direct removal but also by reducing the levels of soluble ubiquitinated proteins inside cells. This finding is supported by studies in animals null for essential autophagic genes in the central nervous system that, even in the absence of a specific pro-aggregating pathogenic protein, display a full aggregopathy-like phenotype (Hara et al. 2006; Komatsu et al. 2006). Aggregates of polyubiquitinated proteins accumulate in the neurons of these animals and result in neurodegenerative symptoms shortly after birth. However, analysis of the content of soluble polyubiquitinated proteins in these animals also revealed a marked increase in this protein pool, leading to the proposal that macroautophagy, likely through its ability to sequester soluble cytosolic cargo "in bulk," contributes to reducing the deposit of these soluble proteins into aggregates (Hara et al. 2006).

CMA also contributes to the elimination of pathogenic proteins but only while still in a soluble state. Although, as indicated above, only proteins bearing a targeting motif in their amino acid sequence can undergo degradation through this pathway, this motif is present in about 30 % of proteins in the soluble proteome

(Dice 1990). In addition, because the motif is based on the properties of the amino acids rather than on the exact sequence, posttranslational modifications such as phosphorylation or acetylation could generate a motif in a protein that does not contain it in the unmodified form (Cuervo 2010). To be recognized for CMA, the targeting motif has to be accessible for the chaperone to bind, which often only happens when some level of unfolding occurs. In other instances, the motif is exposed upon the release of interacting proteins that normally mask this area of the protein. Oxidation of most substrates enhances their degradation through CMA, likely by facilitating access of the chaperone to the targeting motif. Conditions leading to moderate but chronic oxidative stress, such as those that often associate with neurodegeneration, lead to activation of CMA to facilitate the removal of the oxidized proteins (Kiffin et al. 2004). It is not clear yet what determines whether an oxidized protein will be delivered to the ubiquitin-proteasome system or to CMA for degradation.

Other protein modifications have also been shown to facilitate their degradation through CMA. For example, denatured unfolded proteins, truncated proteins or proteins modified after exposure to specific toxic compounds have all been shown to be more readily amenable for CMA targeting and degradation (Cuervo et al. 1998, 1999, 2004; Martinez-Vicente et al. 2008). Although CMA has been shown to contribute to the degradation of the functional form of proteins that, when mutated or modified, become pathogenic, such as α-synuclein or tau, in most instances, as described in detail in the next section, the pathogenic variants can no longer be degraded through this pathway. An important difference with CMA compared to the other autophagic pathways is the need for cargo to undergo complete unfolding before it can be translocated across the lysosomal membrane (Salvador et al. 2000). Consequently, once pathogenic proteins organize into irreversible oligomeric variants, they can no longer gain access to the lysosomal lumen via CMA. Of therapeutic interest is the fact that recent studies have shown that recruitment of CMA for the removal of proteins that are not usually cleared by this pathway is possible (Bauer et al. 2010). For example, although full-length huntingtin protein contains several CMA-targeting motifs, they are not normally utilized for clearance of this protein, whose degradation through CMA under normal conditions is negligible. Artificial expression in cells and mouse models of huntingtin-interacting proteins bearing additional CMA-targeting motifs has been shown to be effective in ameliorating the disease in animal models by facilitating CMA of mutant huntingtin.

Autophagy as a target of proteotoxicity

Lysosomes and the autophagic machinery are not spared the toxicity associated with the presence of pathogenic proteins inside cells. These proteins can become toxic tools secondary to injuries such as acute oxidative damage or other conditions leading to the generation of reactive oxygen species. These species and the proteins modified by them can damage integral components of the lysosomal membrane, with the subsequent leakage of hydrolases to the cytosol and a compromise in the clearance of autophagosomes (Terman 1995). Furthermore, even in the absence of external injury, certain proteins can compromise the normal functioning of the autophagic/lysosomal system; in fact, their direct negative impact on this quality control mechanism underlies the basis of several common neurodegenerative disorders.

Alterations in macroautophagy can occur at very different levels, and understanding the step(s) compromised in this process is important in order to design targeted interventions in the future (Wong and Cuervo 2010). Defects in the formation of autophagosomes can originate from both changes in the Atgs that participate in this step and enhanced activation of inhibitory regulators of this process, such as insulin signaling and the mTOR complex. Different Atgs have been described to co-aggregate with pathogenic proteins in a non-specific manner, reducing in this way their availability to engage in the autophagic pathway. Post-translational modifications, such as phosphorylation and acetylation, of some of the essential Atgs modulate their activity (Lee et al. 2008; Grotemeier et al. 2010); consequently, conditions resulting in changes in the modification effectors could, at least in theory, compromise autophagosome formation (Winslow et al. 2010). Studies with transgenic animal knock-outs for essential autophagic genes in different tissues have revealed that accumulation of ubiquitin-positive aggregates in the cytosol of the affected cells is a feature common to most of these models (Komatsu et al. 2005, 2006; Hara et al. 2006; Nakai et al. 2007). Consequently, pathogenic proteins that interfere with these very early steps of the autophagic process will result in further alterations in protein homeostasis and cellular toxicity. Interventions that enhance autophagosome formation should thus be beneficial in these settings.

There are other instances in which pathogenic proteins do not affect nucleation, formation of the limiting membrane or self-sealing into an autophagosome but do interfere with the ability of the forming autophagosome to recognize cargo. For example, the presence of apparently "empty" autophagosomes has been observed in animal models and peripheral lymphoblasts from patients with Huntington's disease (Martinez-Vicente et al. 2010). The abnormal interaction of the mutant protein in this disease with the inner part of the autophagosome membrane as it forms seems to prevent the selective recognition of cargo. This failure could explain in part the higher content of intracellular organelles in these cells. Although further studies are required, the tight interaction between the pathogenic huntingtin protein and some of the macroautophagic cargo-recognizing molecules suggests that trapping of these proteins in the inner part of the autophagosome may prevent their selective binding to the cargo (Martinez-Vicente et al. 2010). Because the rest of the autophagic steps are for the most part preserved in this disease, autophagosomes can progress to fuse with lysosomes and degrade any cargo internalized even randomly in their lumen. Consequently, upregulation of autophagosome formation should help increase organelle turnover; in fact, early studies have shown that treatment with rapamycin, which upregulates macroautophagy by inhibiting mTOR, reduces cell death and improves cellular homeostasis in experimental models of Huntington's disease (Ravikumar et al. 2004).

In theory, any pathogenic protein that affects vesicular trafficking could also have a negative impact on macroautophagy, because autophagosomes move along microtubules to reach their terminal compartment (Webb et al. 2004; Iwata et al. 2005; Kochl et al. 2006). Defects could result from the direct interaction of the pathogenic protein with the cytoskeleton network, motor proteins or proteins on the surface of the vesicles that facilitate their anchoring to the motor proteins. Recent studies have also provided examples of pathogenic proteins that, by interacting with components at the surface of autophagosomes or lysosomes, prevent the fusion between both compartments (Lee et al. 2010b).

The lysosomes contribute to macroautophagy the hydrolases required for molecular breakdown. In addition, fusion of the autophagosome with lysosomes also results in acidification of the autophagosome lumen, thus providing an ideal milieu for degradation. Consequently, toxic proteins that compromise lysosomal function by modifying their acidification capabilities or by inducing loss of stability of the lysosomal membrane indirectly interfere with macroautophagy. In this respect, most lysosomal storage disorders in which undigested metabolites accumulate in lysosomes will also have an associated compromise in macroautophagy. In addition, examples of pathogenic proteins that directly interfere with lysosomal function in the context of protein conformational disorders have started to appear. Presenilin 1, one of the proteins mutated in certain forms of AD, has been recently linked to the altered lysosomal pH observed in the affected neurons in cellular and animal models (Lee et al. 2010a). The molecular basis of the deficient acidification lies in a novel function for presenilin 1, as the chaperone that facilitates proper glycosylation of components of the lysosomal proton pump in the ER (Lee et al. 2010a). The massive accumulation of autophagosomes observed in all those conditions in which toxic proteins interfere with the last steps in macroautophagy (those that follow autophagosomal formation) makes it unlikely that induction of autophagy will work in these circumstances. Instead, interventions should be targeted at directly correcting the defect in autophagosome clearance, be it at the vesicular trafficking level, in the direct fusion between the membranes of both vesicular compartments or at the level of cargo breakdown.

Alterations in CMA secondary to the action of pathogenic proteins have been recently described in familial forms of PD and in some tauopathies. Pathogenic mutations in α-synuclein do not alter the recognition of these molecules by hsc70 in the cytosol, but instead they compromise their lysosomal translocation via CMA (Cuervo et al. 2004; Martinez-Vicente et al. 2008). Both mutant forms of α-synuclein as well as post-translationally modified forms of the wild type protein can be delivered to lysosomes by hsc70, where they bind with abnormally high affinity to LAMP-2A. For reasons yet unclear, these pathogenic proteins fail to translocate across the lysosomal membrane, but because they remain tightly bound to the CMA receptor, they compromise clearance of any other protein through this pathway (Cuervo et al. 2004; Martinez-Vicente et al. 2008). A similar CMA shut-down has been proposed to be induced by other PD mutant proteins, such as ubiquitin carboxyl-terminal hydrolase L1 (UCH-L1; Kabuta et al. 2008), and even beyond PD, with the identification that specific mutant forms of tau – a protein linked to AD pathogenesis – also interact in an abnormal way with components of CMA at the lysosomal membrane and block the degradation of other substrates (Wang et al. 2009). The consequences of the binding of pathogenic tau are more detrimental for lysosomes than those of PD-related proteins, because tau fragments organize in oligomeric structures that compromise the integrity of the lysosomal membrane (Wang et al. 2009). As in the case of macroautophagy, because the lysosome is the terminal compartment directly responsible for the breakdown of CMA substrates, any toxic protein that interferes with lysosomal acidification, hydrolases content or membrane stability will also lead to compromised CMA activity.

Concluding remarks

The dual interplay between pathogenic proteins and the different autophagic pathways described in this chapter could explain the chronic and perpetuating nature of most protein conformational disorders. Autophagy plays a very active role in the continuous elimination of pathogenic proteins as part of the global cellular defense against proteotoxicity. Consequently, when specific pathogenic proteins target the autophagic systems and interfere with their regular function, a vicious circle is often created in which the inability to eliminate the toxic proteins will further accentuate the defect on autophagic clearance. In addition, other factors and conditions that lead to reduced autophagic activity, such as acute stressors, states of chronic oxidative stress or aging, can thus act as aggravating factors in protein conformational disorders. Understanding if the autophagic defect observed in different types of neurodegenerations is primary or secondary to the effect of pathogenic proteins in these systems and deciphering the specific step(s) affected by these proteins are essential requirements for any future therapeutic intervention based on manipulations of the autophagic system.

Acknowledgments Work in our laboratory is supported by NIH grants from NIA (AG021904, AG031782), NIDDK (DK041918), NINDS (NS038370), a Glenn Foundation Award and a Hirsch/Weill-Caulier Career Scientist Award. Susmita Kaushik is supported by NIA T32AG023475.

References

Arias E, Cuervo AM (2011) Chaperone-mediated autophagy in protein quality control. Curr Opin Cell Biol 23:184–189

Asanuma K, Tanida I, Shirato I, Ueno T, Takahara H, Nishitani T, Kominami E, Tomino Y (2003) MAP-LC3, a promising autophagosomal marker, is processed during the differentiation and recovery of podocytes from PAN nephrosis. FASEB J 17:1165–1167

Balch WE, Morimoto RI, Dillin A, Kelly JW (2008) Adapting proteostasis for disease intervention. Science 319:916–919

Bandyopadhyay U, Kaushik S, Varticovski L, Cuervo AM (2008) The chaperone-mediated autophagy receptor organizes in dynamic protein complexes at the lysosomal membrane. Mol Cell Biol 28:5747–5763

Bandyopadhyay U, Sridhar S, Kaushik S, Kiffin R, Cuervo AM (2010) Identification of Regulators of Chaperone-Mediated Autophagy. Mol Cell 39:535–547

Bauer PO, Goswami A, Wong HK, Okuno M, Kurosawa M, Yamada M, Miyazaki H, Matsumoto G, Kino Y, Nagai Y, Nukina N (2010) Harnessing chaperone-mediated autophagy for the selective degradation of mutant huntingtin protein. Nat Biotechnol 28:256–263

Chiang H, Terlecky S, Plant C, Dice JF (1989) A role for a 70-kilodalton heat shock protein in lysosomal degradation of intracellular proteins. Science 246:382–385

Cuervo AM (2010) Chaperone-mediated autophagy: selectivity pays off. Trends Endocrin Met 21:142–150

Cuervo AM, Hu W, Lim B, Dice JF (1998) IkB is a substrate for a selective pathway of lysosomal proteolysis. Mol Biol Cell 9:1995–2010

Cuervo AM, Hildebrand H, Bomhard EM, Dice JF (1999) Direct lysosomal uptake of alpha2-microglobulin contributes to chemically induced nephropathy. Kidney Int 55:529–545

Cuervo AM, Stefanis L, Fredenburg R, Lansbury PT, Sulzer D (2004) Impaired degradation of mutant alpha-synuclein by chaperone-mediated autophagy. Science 305:1292–1295

Dice JF (1990) Peptide sequences that target cytosolic proteins for lysosomal proteolysis. Trends Biochem Sci 15:305–309

Dice JF (2007) Chaperone-mediated autophagy. Autophagy 3:295–299

Dikic I, Johansen T, Kirkin V (2010) Selective autophagy in cancer development and therapy. Cancer Res 70:3431–3434

Dubouloz F, Deloche O, Wanke V, Cameroni E, De Virgilio C (2005) The TOR and EGO protein complexes orchestrate microautophagy in yeast. Mol Cell 19:15–26

Grotemeier A, Alers S, Pfisterer SG, Paasch F, Daubrawa M, Dieterle A, Viollet B, Wesselborg S, Proikas-Cezanne T, Stork B (2010) AMPK-independent induction of autophagy by cytosolic Ca2+ increase. Cell Signal 22:914–925

Hara T, Nakamura K, Matsui M, Yamamoto A, Nakahara Y, Suzuki-Migishima R, Yokoyama M, Mishima K, Saito I, Okano H, Mizushima N (2006) Suppression of basal autophagy in neural cells causes neurodegenerative disease in mice. Nature 441:885–889

Iwata A, Riley BE, Johnston JA, Kopito RR (2005) HDAC6 and microtubules are required for autophagic degradation of aggregated huntingtin. J Biol Chem 280:40282–40292

Kabuta T, Furuta A, Aoki A, Furuta K, Wada K (2008) Aberrant interaction between Parkinson disease-associated mutant UCH-L1 and the lysosomal receptor for chaperone-mediated autophagy. J Biol Chem 283:23731–23738

Kiffin R, Christian C, Knecht E, Cuervo A (2004) Activation of chaperone-mediated autophagy during oxidative stress. Mol Biol Cell 15:4829–4840

Kirkin V, McEwan DG, Novak I, Dikic I (2009) A role for ubiquitin in selective autophagy. Mol Cell 34:259–269

Kochl R, Hu X, Chan E, Tooze S (2006) Microtubules facilitate autophagosome formation and fusion of autophagosomes with endosomes. Traffic 7:129–145

Komatsu M, Waguri S, Ueno T, Iwata J, Murata S, Tanida I, Ezaki J, Mizushima N, Ohsumi Y, Uchiyama Y, Kominami E, Tanaka K, Chiba T (2005) Impairment of starvation-induced and constitutive autophagy in Atg7-deficient mice. J Cell Biol 169:425–434

Komatsu M, Waguri S, Chiba T, Murata S, Iwata J, Tanida I, Ueno T, Koike M, Uchiyama Y, Kominami E, Tanaka K (2006) Loss of autophagy in the central nervous system causes neurodegeneration in mice. Nature 441:880–884

Lamark T, Kirkin V, Dikic I, Johansen T (2009) NBR1 and p62 as cargo receptors for selective autophagy of ubiquitinated targets. Cell Cycle 8:1986–1990

Lee IH, Cao L, Mostoslavsky R, Lombard DB, Liu J, Bruns NE, Tsokos M, Alt FW, Finkel T (2008) A role for the NAD-dependent deacetylase Sirt1 in the regulation of autophagy. Proc Natl Acad Sci USA 105:3374–3379

Lee J-H, Yu W, Kumar A, Lee S, Mohan P, Peterhoff C, Wolfe D, Martinez-Vicente M, Massey A, Sovak G, Uchiyama Y, Westaway D, Sisodia S, Cuervo AM, Nixon RA (2010a) PS1 mutations in Alzheimer's disease disrupt lysosomal proteolysis and autophagy. Cell 7:1146–1158

Lee JY, Koga H, Kawaguchi Y, Tang W, Wong E, Gao YS, Pandey UB, Kaushik S, Tresse E, Lu J, Taylor JP, Cuervo AM, Yao TP (2010b) HDAC6 controls autophagosome maturation essential for ubiquitin-selective quality-control autophagy. EMBO J 29:969–980

Martinez-Vicente M, Talloczy Z, Kaushik S, Massey A, Mazzulli J, Mosharov E, Hodara R, Fredenburg R, Wu D, Follenzi A, Dauer W, Przedborski S, Ischiropoulos H, Lansbury P, Sulzer D, Cuervo A (2008) Dopamine-modified alpha-synuclein blocks chaperone-mediated autophagy. J Clin Invest 118:777–788

Martinez-Vicente M, Talloczy Z, Wong E, Tang GM, Koga H, Kaushik S, de Vries R, Arias E, Harris S, Sulzer D, Cuervo AM (2010) Cargo recognition failure is responsible for inefficient autophagy in Huntington's disease. Nat Neurosci 13:567–574

Marzella L, Ahlberg J, Glaumann H (1981) Autophagy, heterophagy, microautophagy and crinophagy as the means for intracellular degradation. Virchows Archiv B Cell Pathol 36:219–234

Mizushima N, Levine B, Cuervo AM, Klionsky DJ (2008) Autophagy fights disease through cellular self-digestion. Nature 451:1069–1075

Morimoto RI (2008) Proteotoxic stress and inducible chaperone networks in neurodegenerative disease and aging. Genes Dev 22:1427–1438

Morimoto RI, Cuervo AM (2009) Protein homeostasis and aging: taking care of proteins from the cradle to the grave. J Gerontol A Biol Sci Med Sci 64:167–170

Nakai A, Yamaguchi O, Takeda T, Higuchi Y, Hikoso S, Taniike M, Omiya S, Mizote I, Matsumura Y, Asahi M, Nishida K, Hori M, Mizushima N, Otsu K (2007) The role of autophagy in cariomyocytes in the basal state and in response to hemodynamic stress. Nat Med 13:619–624

Noda NN, Ohsumi Y, Inagaki F (2010) Atg8-family interacting motif crucial for selective autophagy. FEBS Lett 584:1379–1385

Novak I, Dikic I (2011) Autophagy receptors in developmental clearance of mitochondria. Autophagy 7:301–303

Prahlad V, Cornelius T, Morimoto RI (2008) Regulation of the cellular heat shock response in Caenorhabditis elegans by thermosensory neurons. Science 320:811–814

Ravikumar B, Duden R, Rubinsztein DC (2002) Aggregate-prone proteins with polyglutamine and polyalanine expansions are degraded by autophagy. Hum Mol Genet 11:1107–1117

Ravikumar B, Vacher C, Berger Z, Davies JE, Luo S, Oroz LG, Scaravilli F, Easton DF, Duden R, O'Kane CJ, Rubinsztein DC (2004) Inhibition of mTOR induces autophagy and reduces toxicity of polyglutamine expansions in fly and mouse models of Huntington disease. Nat Genet 36:585–595

Roberts P, Moshitch-Moshkovitz S, Kvam E, O'Toole E, Winey M, Goldfarb D (2003) Piecemeal microautophagy of nucleus in Saccharomyces cerevisiae. Mol Biol Cell 14:129–141

Sahu R, Kaushik S, Cannizzo E, Scharf B, Follenzi A, Clement C, Potolicchio I, Nieves E, Cuervo A, Santambrogio L (2011) Microautohagy of cytosolic proteins by late endosomes. Dev Cell 20:131–139

Sakai Y, Koller A, Rangell L, Keller G, Subramani S (1998) Peroxisome degradation by microautophagy in *Pichia pastoris*. Identification of specific steps and morphological intermediates. J Cell Biol 141:625–636

Salvador N, Aguado C, Horst M, Knecht E (2000) Import of a cytosolic protein into lysosomes by chaperone-mediated autophagy depends on its folding state. J Biol Chem 275:27447–27456

Sarkar S, Davies JE, Huang Z, Tunnacliffe A, Rubinsztein DC (2007) Trehalose, a novel mTOR-independent autophagy enhancer, accelerates the clearance of mutant huntingtin and alpha-synuclein. J Biol Chem 282:5641–5652

Sarkar S, Ravikumar B, Rubinsztein DC (2009) Autophagic clearance of aggregate-prone proteins associated with neurodegeneration. Meth Enzymol 453:83–110

Terman A (1995) The effect of age on formation and elimination of autophagic vacuoles in mouse hepatocytes. Gerontology 41:319–325

Wang Y, Martinez-Vicente M, Kruger U, Kaushik S, Wong E, Mandelkow EM, Cuervo AM, Mandelkow E (2009) Tau fragmentation, aggregation and clearance: the dual role of lysosomal processing. Hum Mol Genet 18:4153–4170

Webb JL, Ravikumar B, Rubinsztein DC (2004) Microtubule disruption inhibits autophagosome-lysosome fusion: implications for studying the roles of aggresomes in polyglutamine diseases. Int J Biochem Cell Biol 36:2541–2550

Winslow AR, Chen CW, Corrochano S, Acevedo-Arozena A, Gordon DE, Peden AA, Lichtenberg M, Menzies FM, Ravikumar B, Imarisio S, Brown S, O'Kane CJ, Rubinsztein DC (2010) alpha-Synuclein impairs macroautophagy: implications for Parkinson's disease. J Cell Biol 190:1023–1037

Wong E, Cuervo AM (2010) Autophagy gone awry in neurodegenerative diseases. Nat Neurosci 13:805–811

Wong ES, Tan JM, Soong WE, Hussein K, Nukina N, Dawson VL, Dawson TM, Cuervo AM, Lim KL (2008) Autophagy-mediated clearance of aggresomes is not a universal phenomenon. Hum Mol Genet 17:2570–2582

Yamamoto A, Cremona M, Rothman J (2006) Autophagy-mediated clearance of huntingtin aggregates triggered by the insulin-signaling pathway. J Cell Biol 172:719–731

Yang Z, Klionsky DJ (2009) An overview of the molecular mechanism of autophagy. Curr Top Microbiol Immunol 335:1–32

Yang Z, Klionsky DJ (2010a) Eaten alive: a history of macroautophagy. Nat Cell Biol 12:814–822

Yang Z, Klionsky DJ (2010b) Mammalian autophagy: core molecular machinery and signaling regulation. Curr Opin Cell Biol 22:124–131

Yuan W, Tuttle DL, Shi YJ, Ralph GS, Dunn WA Jr (1997) Glucose-induced microautophagy in Pichia pastoris requires the alpha-subunit of phosphofructokinase. Journal of Cell Science 110:1935–1945

Quality control of proteins and organelles by autophagy

Noboru Mizushima

Abstract Macroautophagy is one of the major degradation pathways in the cell. In autophagy, intracellular components are sequestered by autophagosomes and then degraded upon fusion with lysosomes. Genetic analyses of autophagy in mammals have revealed that autophagy is important not only for an adaptive response to starvation but also for quality control of intracellular proteins and organelles. These roles of autophagy are tightly connected to various physiological and pathological processes such as neurodegeneration and tumorigenesis.

Introduction

Macroautophagy (which we will simply refer to as autophagy) is an intracellular degradation system mediated by the autophagosome. A small portion of the cytoplasm is engulfed by an isolation membrane/phagophore, which results in the formation of an autophagosome. The outer membrane of the autophagosome fuses with the lysosome, and materials inside the autophagosome are degraded by the lysosomal hydrolases (Fig. 1). To date, yeast genetic studies have identified more than 30 autophagy-related (*ATG*) genes, many of which are conserved in higher eukaryotes (Suzuki and Ohsumi 2007; Xie and Klionsky 2007; Longatti and Tooze 2009). Reverse genetic approaches performed in various organisms, including mice, have revealed that autophagy is involved in many physiological processes, such as maintenance of the amino acid pool during starvation and preimplantation embryo development, intracellular quality control, tumor suppression, antigen presentation and killing of intracellular microorganisms (Rubinsztein 2006; Cecconi and Levine 2008; Mizushima et al. 2008; Mizushima and Levine

N. Mizushima (✉)
Department of Physiology and Cell Biology, Tokyo Medical and Dental University, 1-5-45 Yushima, Bunkyo-ku, Tokyo 113-8519, Japan
e-mail: nmizu.phy2@tmd.ac.jp

R.I. Morimoto and Y. Christen (eds.), *Protein Quality Control in Neurodegenerative Diseases*, Research and Perspectives in Alzheimer's Disease,
DOI 10.1007/978-3-642-27928-7_6,

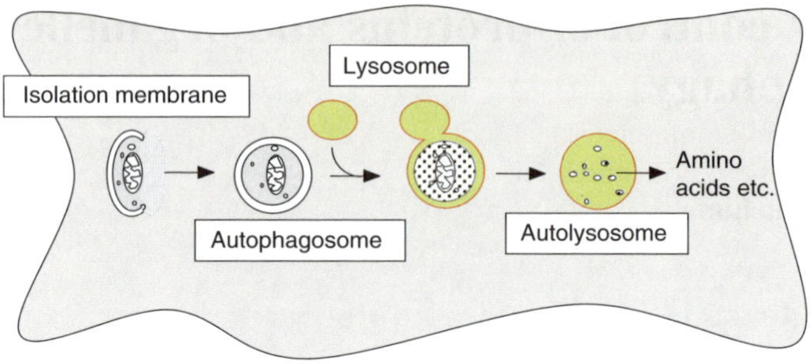

Fig. 1 The process of macroautophagy in mammalian cells. A portion of cytoplasm, including organelles, is enclosed by an isolation membrane (also called a phagophore) to form an autophagosome. The outer membrane of the autophagosome fuses with the lysosome and the internal material is degraded. The resultant degradation products can be used for many different purposes, such as new protein synthesis, energy production and gluconeogenesis

2010; Levine et al. 2011). In this review, the role of autophagy as an intracellular quality control system will be a focus, and its potential relationship with neurodegeneration and cancer will be discussed.

Protein quality control by autophagy

The evolutionarily conserved role of autophagy is an adaptation response to stress conditions, particularly nutrient starvation. Yeast cells, cultured mammalian cells and even mice (whole organism) have the capacity to up-regulate autophagy following starvation (Tsukada and Ohsumi 1993; Kuma et al. 2004).

On the other hand, "constitutive autophagy" or "basal autophagy" is an important quality control system, particularly in the nervous system and liver. The first demonstration of this role came from analysis of mice with liver-specific deletion of *Atg7*, a gene essential for autophagosome formation (Komatsu et al. 2005). These mice developed severe hepatomegaly with intracellular accumulation of ubiquitin-positive aggregates and deformed organelles.

The importance of autophagy as an intracellular clearance system was further confirmed by analysis of mice with neural cell-specific deletion of either *Atg7* (*Atg7*$^{flox/flox}$;*Nestin-Cre*; Komatsu et al. 2006) or Atg5, another gene required for autophagosome formation (*Atg5*$^{flox/flox}$;*Nestin-Cre*; Hara et al. 2006). These mice developed progressive motor deficits after three weeks of age, such as ataxic gait, impaired motor coordination, abnormal limb clasping reflexes and systemic tremor. Some mice sporadically died after three weeks of age. Degenerative changes of in neurons were observed, such as partial loss of Purkinje cells and cerebral pyramidal cells and axonal swelling. Like liver-specific Atg7 knockout mice, neurons of *Atg5*$^{flox/flox}$;*Nestin-Cre* and *Atg7*$^{flox/flox}$;*Nestin-Cre* mice possessed ubiquitin-positive protein aggregates (inclusion bodies). Systemic deletion of *Atg5* caused neonatal lethal, ubiquitin-positive aggregates to accumulate in many neonatal tissues, such as in hepatocytes, a subset of neurons, the anterior lobe of pituitary gland and the adrenal gland (Hara et al. 2006). These studies suggest that autophagy is critical for cellular homeostasis, even if any disease-associated mutant (aggregate prone) proteins are not expressed.

Selective degradation of p62 by autophagy

Although autophagy has been considered to be a non-selective, bulk process, recent studies have revealed that the autophagosome membrane can selectively recognize certain specific proteins and organelles (Kirkin et al. 2009; Kraft et al. 2010). The best-known autophagy-specific substrate in mammals is p62/SQSTM1 (Bjørkøy et al. 2005; Pankiv et al. 2007). p62 has multiple functions in bone metabolism, obesity, and inflammatory signal transduction (Wooten et al. 2006; Seibenhener et al. 2007; Moscat and Diaz-Meco 2009). p62 has a short motif that interacts with LC3, an autophagosomal protein; therefore, p62 is preferentially degraded by autophagy. Indeed p62 massively accumulates in autophagy-deficient tissues and cells (Bjørkøy et al. 2005; Komatsu et al. 2007).

Autophagic degradation of p62 is physiologically important. p62 (Fig. 2) is a core molecule that forms inclusion bodies and is indeed found in the ubiquitin-positive aggregates that are formed in liver- and neuron-specific, autophagy-deficient mice (Komatsu et al. 2007). When p62 is simultaneously deleted in these Atg knockout mice, the ubiquitin-positive aggregates no longer accumulate. Thus, formation of the protein aggregates can be primarily caused by p62 overexpression (Komatsu et al. 2007). Accordingly, liver phenotypes such as hepatomegaly and liver dysfunction are dramatically improved in the $Atg7^{-/-}p62^{-/-}$ double knockout mice (Komatsu et al. 2007). The cytotoxic role of p62 has been further clarified. p62 interacts with Keap1, a component of a ubiquitin ligase complex for the transcription factor, Nrf2, and inhibits degradation of Nrf2 (Komatsu et al. 2010; Lau et al. 2010). Thus, overexpression of p62 caused by autophagy defects results in stabilization of Nrf2, which leads to transcriptional activation of Nrf2 target genes, such as antioxidant proteins and detoxification enzymes. This hyperactivation of the Nrf2 pathway is toxic, at least in the liver, because liver damage in *Atg7* knockout mice can be ameliorated by simultaneous deletion of *Nrf2* and further exacerbated by deletion of *Keap1* (Komatsu et al. 2010).

On the other hand, neurological symptoms of neural cell-specific *Atg7* knockout mice are not ameliorated by deletion of p62, although ubiquitin-positive aggregates disappear in neurons (Komatsu et al. 2007). Therefore, autophagy could be important for more general intracellular quality control, not only for turnover of p62.

It has also been hypothesized that p62 functions as an adaptor or cargo receptor for ubiquitinated proteins because p62 has a ubiquitin-associated (UBA) domain (Kim et al. 2008; Kirkin et al. 2009). However, this model has not been experimentally proved in mouse models. Deletion of p62 in *Atg7* knockout mice does not

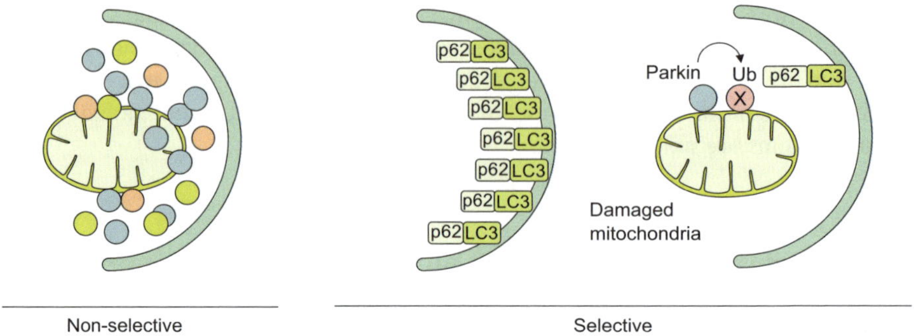

Fig. 2 Bulk versus selective autophagy. Most autophagic substrates can be non-selectively engulfed by the autophagosome (*left*). p62 and other selective substrates have the Atg8/LC3 interacting motif and are recognized by autophagosomal membranes for preferential degradation (*middle*). Depolarized mitochondria can also be selectively degraded by autophagy, although it remains unknown how mitochondria are recognized (*right*). Ub, poly-ubiquitin chain

significantly reduce the amount of ubiquitinated proteins (Komatsu et al. 2010). Another recent study suggests that accumulation of ubiquitinated proteins in autophagy-deficient hepatocytes and neurons could be a secondary consequence of Nrf2 activation, rather than a direct result of a defect in clearance of ubiquitinated proteins (Riley et al. 2010).

p62 is recruited to the autophagosome formation site

Hierarchical relationships among Atg and Atg-related proteins have been analyzed in both yeast (Suzuki et al. 2007) and mammals (Itakura and Mizushima 2010). These studies suggest that the Atg1/ULK1 protein complex is the most upstream unit and is followed by the Atg14/PI3-kinase complex, the Atg12–Atg5–Atg16L1 complex and Atg8/LC3–PE. These proteins accumulate at a perivacuolar structure called the preautophagosomal structure (PAS) in yeast. In mammals, these proteins are recruited to some domains on the endoplasmic reticulum (ER), which may be equivalent to the yeast PAS (Itakura and Mizushima 2010; Matsunaga et al. 2010).

With this hierarchical map of mammalian Atg proteins, we analyzed at which step the selective autophagy substrate p62 is recruited to autophagic structures. Although it is generally thought that p62 is recruited to preexisting isolation membrane in an LC3-dependent manner (Kirkin et al. 2009; Lamark et al. 2009), we found that p62 and its related protein NBR1 could localize to the ER-associated autophagosome formation site independently of LC3 localization to membranes (Itakura and Mizushima 2011). p62 co-localized with upstream autophagy factors such as ULK1 even when autophagosome formation was blocked by wortmannin treatment or genetic ablation of FIP200. Although p62 is not essential for autophagy (Komatsu et al. 2007), it is still possible that p62 has the ability to recruit Atg proteins and to determine where autophagosome should be generated.

As p62 has been suggested to also be important for pexophagy (Kim et al. 2008) and elimination of bacteria (Yoshikawa et al. 2009), p62 may play a general role in recruiting Atg proteins to initiate autophagosome formation. The p62–LC3 interaction might be important for proper incorporation of p62 inside autophagosomes and/or recruitment of additional p62 into autophagosomes.

Quality control of mitochondria by autophagy and its implication in the pathogenesis of Parkinson's disease

Unlike the ubiquitin-proteasome system, autophagy can degrade entire organelles as well as cytoplamic proteins. The role of autophagy in mitochondrial quality control is now a hot topic in this field, particularly in its potential involvement in the pathogenesis of Parkinson disease (PD). PD is a progressive neurodegenerative disease that is characterized by postural changes, resting tremor, muscle rigidity, and weakness, symptoms that are primarily caused by loss of dopaminergic neurons in the substantia nigra. Many of the PD-related-gene products, such as Parkin, PINK1, DJ-1 and OMI, have been suggested to link to mitochondria (Abou-Sleiman et al. 2006; Zhu and Chu 2010).

Parkin is an E3 ubiquitin ligase and its mutation causes autosomal recessive juvenile PD (Kitada et al. 1998). Recent studies have revealed that Parkin is important for elimination of damaged mitochondria by autophagy, which is termed "mitophagy" (Youle and Narendra 2011). Narendra et al. (2008) first demonstrated that Parkin translocates from the cytosol to depolarized mitochondria, and triggers mitophagy. Targeting of Parkin to mitochondria requires PTEN-induced putative kinase 1 (Pink1), another PD-associated gene product (Gegg et al. 2010; Geisler et al. 2010; Kawajiri et al. 2010; Matsuda et al. 2010; Narendra et al. w2010b; Rakovic et al. 2010; Vives-Bauza et al. 2010; Ziviani et al. 2010). Although Pink1 is an extremely unstable mitochondrial protein due to continuous degradation by Parl, it becomes stabilized upon mitochondrial depolarization and subsequently recruits Parkin (Jin et al. 2010; Shi et al. 2011).

However, how Parkin induces mitophagy has not been fully elucidated. To date, several mitochondrial proteins — voltage-dependent anion channel 1 (VDAC1; Geisler et al. 2010), mitofusin (a mitochondrial pro-fusion factor; Gegg et al. 2010; Poole et al. 2010; Tanaka et al. 2010; Ziviani et al. 2010), Bcl-2 (Chen et al. 2010), and Drp1 (Wang et al. 2011) — have been shown to be ubiquitinated by Parkin. These ubiquitinated proteins may recruit the autophagy adaptor p62 (Ding et al. 2010; Geisler et al. 2010; Lee et al. 2010). However, the role of p62 remains controversial; p62 seems to be essential for mitochondrial clustering at the perinuclear region, but not for mitophagy itself (Narendra et al. 2010a; Okatsu et al. 2010). We also found that LC3-p62 interaction is not the first recognition step of damaged mitochondria, and upstream Atg proteins can be recruited in a LC3-independent manner (Itakura et al. 2012). Yeast mitochondria have the cargo receptor, Atg32, which interacts with Cvt11 as well as Atg8 (Kanki et al. 2009; Okamoto et al. 2009). However, its counterpart seems not to be present in mammals.

More recently, several reports revealed that Parkin is also important for proteasomal degradation of a wide range of outer mitochondrial membrane proteins (Tanaka et al. 2010; Chan et al. 2011; Yoshii et al. 2011). In Parkin-overexpressing cells, mitochondrial depolarization induces proteasome-dependent degradation of mitochondrial proteins of the outer membrane and intermembrane space, whereas those of the inner membrane and matrix are mainly degraded by mitophagy (Chan et al. 2011; Yoshii et al. 2011). Proteasome and p97/VCP are recruited to these damaged mitochondria (Tanaka et al. 2010; Chan et al. 2011; Yoshii et al. 2011). Electron microscopy revealed rupture of the outer membrane in most of these mitochondria (Yoshii et al. 2011). However, whether or not the proteasomal degradation of outer membrane proteins is required for mitophagy remains controversial. We observed that inhibition of proteasome does not affect mitophagy, suggesting that Parkin has two independent functions: turnover of outer membrane proteins at a molecular level and turnover of entire mitochondria at an organelle level (Yoshii et al. 2011). On the other hand, two other groups found that the degradation of the outer membrane proteins contributes to mitophagy (Tanaka et al. 2010; Chan et al. 2011). Ubiquitination and degradation of mitofusin may affect mitochondrial fusion, which would facilitate mitophagy, even though mitofusin itself is not essential for mitophagy (Tanaka et al. 2010).

Autophagy and tumorigenesis

It has been suggested that autophagy is also involved in tumor suppression (Levine and Kroemer 2008; Chen and Debnath 2010; White et al. 2010). Using cell culture and allografted tumor models, several tumor-suppressive mechanisms have been proposed: 1) suppression of tumorigenic inflammation (Degenhardt et al. 2006), 2) mitigation of metabolic stress and genomic damage (Karantza-Wadsworth et al. 2007; Mathew et al. 2007), and 3) p62 (also called SQSTM1) degradation (Mathew et al., 2009). However, results from in vivo models have been limited. Beclin 1 heterozygous mutant mice showed an increased frequency of spontaneous cancers in the lung, liver and lymphoid tissues (Qu et al. 2003; Yue et al. 2003). Several Beclin 1-interacting proteins, including Ambra 1 (Fimia et al. 2007), Bif-1 (Takahashi et al. 2007), and UVRAG (Liang et al. 2006), also have tumor suppressive or anti-proliferation effects. However, these factors are not strictly specific for autophagy (Funderburk et al. 2010; He and Levine 2010; Thoresen et al. 2010).

The tumor-suppressive role of autophagy was further confirmed using models deficient for more downstream-specific autophagy factors. Since systemic deletion of *Atg5* causes neonatal lethality (Kuma et al. 2004), we generated a new mouse model with systemic mosaic deletion of Atg5 (Takamura et al. 2011). These mice developed multiple benign tumors only in the liver, but not in other tissues. Swollen mitochondria and oxidative stress and genomic damage responses were detected in the hepatic tumor cells.

Liver-specific, *Atg7*-deficient mice also developed liver tumors (Takamura et al. 2011). Importantly, concomitant deletion of p62 reduced the size of the liver tumors, although it did not completely suppress tumor formation (Takamura et al. 2011). Thus, accumulation of p62 caused by autophagy deficiency may contribute to tumor progression rather than tumorigenesis. These results suggest that turnover of both proteins and organelles is critically important for prevention of spontaneous tumorigenesis and subsequent tumor progression.

Acknowledgment This work was supported in part by Grants-in-Aid for Scientific Research from the Ministry of Education, Culture, Sports, Science and Technology, Japan, the Takeda Science Foundation, and Funding Program for Next Generation World-Leading Researchers.

References

Abou-Sleiman PM, Muqit MM, Wood NW (2006) Expanding insights of mitochondrial dysfunction in Parkinson's disease. Nat Rev Neurosci 7:207–219

Bjørkøy G, Lamark T, Brech A, Outzen H, Perander M, Øvervatn A, Stenmark H, Johansen T (2005) p62/SQSTM1 forms protein aggregates degraded by autophagy and has a protective effect on huntingtin-induced cell death. J Cell Biol 171:603–614

Cecconi F, Levine B (2008) The role of autophagy in mammalian development: cell makeover rather than cell death. Dev Cell 15:344–357

Chan NC, Salazar AM, Pham AH, Sweredoski MJ, Kolawa NJ, Graham RL, Hess S, Chan DC (2011) Broad activation of the ubiquitin-proteasome system by Parkin is critical for mitophagy. Hum Mol Genet 20:1726–1737

Chen D, Gao F, Li B, Wang H, Xu Y, Zhu C, Wang G (2010) Parkin mono-ubiquitinates BCL-2 and regulates autophagy. J Biol Chem 285:38214–38223

Chen N, Debnath J (2010) Autophagy and tumorigenesis. FEBS Lett 584:1427–1435

Degenhardt K, Mathew R, Beaudoin B, Bray K, Anderson D, Chen G, Mukherjee C, Shi Y, Gelinas C, Fan Y, Nelson DA, Jin S, White E (2006) Autophagy promotes tumor cell survival and restricts necrosis, inflammation, and tumorigenesis. Cancer Cell 10:51–64

Ding WX, Ni HM, Li M, Liao Y, Chen X, Stolz DB, Dorn Ii GW, Yin XM (2010) Nix is critical to two distinct phases of mitophagy: reactive oxygen species (ROS)-mediated autophagy induction and Parkin-ubiqutin-p62-mediated mitochondria priming. J Biol Chem 285:27879–27890

Fimia GM, Stoykova A, Romagnoli A, Giunta L, Di Bartolomeo S, Nardacci R, Corazzari M, Fuoco C, Ucar A, Schwartz P, Gruss P, Piacentini M, Chowdhury K, Cecconi F (2007) Ambra1 regulates autophagy and development of the nervous system. Nature 447:1121–1125

Funderburk SF, Wang QJ, Yue Z (2010) The Beclin 1-VPS34 complex at the crossroads of autophagy and beyond. Trends Cell Biol 20:355–362

Gegg ME, Cooper JM, Chau KY, Rojo M, Schapira AH, Taanman JW (2010) Mitofusin-1 and Mitofusin-2 are ubiquitinated in a PINK1/parkin dependent manner upon induction of mitophagy. Hum Mol Genet 19:4861–4870

Geisler S, Holmstrom KM, Skujat D, Fiesel FC, Rothfuss OC, Kahle PJ, Springer W (2010) PINK1/Parkin-mediated mitophagy is dependent on VDAC1 and p62/SQSTM1. Nat Cell Biol 12:119–131

Hara T, Nakamura K, Matsui M, Yamamoto A, Nakahara Y, Suzuki-Migishima R, Yokoyama M, Mishima K, Saito I, Okano H, Mizushima N (2006) Suppression of basal autophagy in neural cells causes neurodegenerative disease in mice. Nature 441:885–889

He C, Levine B (2010) The Beclin 1 interactome. Curr Opin Cell Biol 22:140–149
Itakura E, Kishi-Itakura C, Koyama-Honda I, Mizushima N (2012) Structures containing Atg9A and the ULK1 complex independently target depolarized mitochondria at initial stages of Parkin-mediated mitophagy. J Cell Sci 125:1488–1499
Itakura E, Mizushima N (2011) p62 targeting to the autophagosome formation site requires self-oligomerization but not LC3 binding. J Cell Biol 192:17–27
Itakura E, Mizushima N (2010) Characterization of autophagosome formation site by a hierarchical analysis of mammalian Atg proteins. Autophagy 6:764–776
Jin SM, Lazarou M, Wang C, Kane LA, Narendra DP, Youle RJ (2010) Mitochondrial membrane potential regulates PINK1 import and proteolytic destabilization by PARL. J Cell Biol 191:933–942
Kanki T, Wang K, Cao Y, Baba M, Klionsky DJ (2009) Atg32 is a mitochondrial protein that confers selectivity during mitophagy. Dev Cell 17:98–109
Karantza-Wadsworth V, Patel S, Kravchuk O, Chen G, Mathew R, Jin S, White E (2007) Autophagy mitigates metabolic stress and genome damage in mammary tumorigenesis. Genes Dev 21:1621–1635
Kawajiri S, Saiki S, Sato S, Sato F, Hatano T, Eguchi H, Hattori N (2010) PINK1 is recruited to mitochondria with parkin and associates with LC3 in mitophagy. FEBS Lett 584:1073–1079
Kim PK, Hailey DW, Mullen RT, Lippincott-Schwartz J (2008) Ubiquitin signals autophagic degradation of cytosolic proteins and peroxisomes. Proc Natl Acad Sci USA 105:20567–20574
Kirkin V, McEwan DG, Novak I, Dikic I (2009) A role for ubiquitin in selective autophagy. Mol Cell 34:259–269
Kitada T, Asakawa S, Hattori N, Matsumine H, Yamamura Y, Minoshima S, Yokochi M, Mizuno Y, Shimizu N (1998) Mutations in the parkin gene cause autosomal recessive juvenile parkinsonism. Nature 392:605–608
Komatsu M, Waguri S, Ueno T, Iwata J, Murata S, Tanida I, Ezaki J, Mizushima N, Ohsumi Y, Uchiyama Y, Kominami E, Tanaka K, Chiba T (2005) Impairment of starvation-induced and constitutive autophagy in Atg7-deficient mice. J Cell Biol 169:425–434
Komatsu M, Waguri S, Chiba T, Murata S, Iwata JI, Tanida I, Ueno T, Koike M, Uchiyama Y, Kominami E, Tanaka K (2006) Loss of autophagy in the central nervous system causes neurodegeneration in mice. Nature 441:880–884
Komatsu M, Waguri S, Koike M, Sou YS, Ueno T, Hara T, Mizushima N, Iwata J, Ezaki J, Murata S, Hamazaki J, Nishito Y, Iemura S, Natsume T, Yanagawa T, Uwayama J, Warabi E, Yoshida H, Ishii T, Kobayashi A, Yamamoto M, Yue Z, Uchiyama Y, Kominami E, Tanaka K (2007) Homeostatic levels of p62 control cytoplasmic inclusion body formation in autophagy-deficient mice. Cell 131:1149–1163
Komatsu M, Kurokawa H, Waguri S, Taguchi K, Kobayashi A, Ichimura Y, Sou YS, Ueno I, Sakamoto A, Tong KI, Kim M, Nishito Y, Iemura S, Natsume T, Ueno T, Kominami E, Motohashi H, Tanaka K, Yamamoto M (2010) The selective autophagy substrate p62 activates the stress responsive transcription factor Nrf2 through inactivation of Keap1. Nat Cell Biol 12:213–223
Kraft C, Peter M, Hofmann K (2010) Selective autophagy: ubiquitin-mediated recognition and beyond. Nat Cell Biol 12:836–841
Kuma A, Hatano M, Matsui M, Yamamoto A, Nakaya H, Yoshimori T, Ohsumi Y, Tokuhisa T, Mizushima N (2004) The role of autophagy during the early neonatal starvation period. Nature 432:1032–1036
Lamark T, Kirkin V, Dikic I, Johansen T (2009) NBR1 and p62 as cargo receptors for selective autophagy of ubiquitinated targets. Cell Cycle 8:1986–1990
Lau A, Wang XJ, Zhao F, Villeneuve NF, Wu T, Jiang T, Sun Z, White E, Zhang DD (2010) A non-canonical mechanism of Nrf2 activation by autophagy deficiency: a direct interaction between Keap1 and p62. Mol Cell Biol 30:3275–3285
Lee JY, Nagano Y, Taylor JP, Lim KL, Yao TP (2010) Disease-causing mutations in parkin impair mitochondrial ubiquitination, aggregation, and HDAC6-dependent mitophagy. J Cell Biol 189:671–679
Levine B, Kroemer G (2008) Autophagy in the pathogenesis of disease. Cell 132:27–42

Levine B, Mizushima N, Virgin HW (2011) Autophagy in immunity and inflammation. Nature 469:323–335

Liang C, Feng P, Ku B, Dotan I, Canaani D, Oh BH, Jung JU (2006) Autophagic and tumour suppressor activity of a novel Beclin1-binding protein UVRAG. Nat Cell Biol 8:688–699

Longatti A, Tooze SA (2009) Vesicular trafficking and autophagosome formation. Cell Death Differ 16:956–965

Mathew R, Kongara S, Beaudoin B, Karp CM, Bray K, Degenhardt K, Chen G, Jin S, White E (2007) Autophagy suppresses tumor progression by limiting chromosomal instability. Genes Dev 21:1367–1381

Mathew R, Karp CM, Beaudoin B, Vuong N, Chen G, Chen HY, Bray K, Reddy A, Bhanot G, Gelinas C, Dipaola RS, Karantza-Wadsworth V, White E (2009) Autophagy suppresses tumorigenesis through elimination of p62. Cell 137:1062–1075

Matsuda N, Sato S, Shiba K, Okatsu K, Saisho K, Gautier CA, Sou YS, Saiki S, Kawajiri S, Sato F, Kimura M, Komatsu M, Hattori N, Tanaka K (2010) PINK1 stabilized by mitochondrial depolarization recruits Parkin to damaged mitochondria and activates latent Parkin for mitophagy. J Cell Biol 189:211–221

Matsunaga K, Morita E, Saitoh T, Akira S, Ktistakis NT, Izumi T, Noda T, Yoshimori T (2010) Autophagy requires endoplasmic reticulum targeting of the PI3-kinase complex via Atg14L. J Cell Biol 190:511–521

Mizushima N, Levine B (2010) Autophagy in mammalian development and differentiation. Nat Cell Biol 12:823–830

Mizushima N, Levine B, Cuervo AM, Klionsky DJ (2008) Autophagy fights disease through cellular self-digestion. Nature 451:1069–1075

Moscat J, Diaz-Meco MT (2009) p62 at the crossroads of autophagy, apoptosis, and cancer. Cell 137:1001–1004

Narendra D, Tanaka A, Suen DF, Youle RJ (2008) Parkin is recruited selectively to impaired mitochondria and promotes their autophagy. J Cell Biol 183:795–803

Narendra DP, Kane LA, Hauser DN, Fearnley IM, Youle RJ (2010a) p62/SQSTM1 is required for Parkin-induced mitochondrial clustering but not mitophagy; VDAC1 is dispensable for both. Autophagy 6:1090–1106

Narendra DP, Jin SM, Tanaka A, Suen DF, Gautier CA, Shen J, Cookson MR, Youle RJ (2010b) PINK1 is selectively stabilized on impaired mitochondria to activate Parkin. PLoS Biol 8: e1000298

Okamoto K, Kondo-Okamoto N, Ohsumi Y (2009) Mitochondria-anchored receptor Atg32 mediates degradation of mitochondria via selective autophagy. Dev Cell 17:87–97

Okatsu K, Saisho K, Shimanuki M, Nakada K, Shitara H, Sou YS, Kimura M, Sato S, Hattori N, Komatsu M, Tanaka K, Matsuda N (2010) p62/SQSTM1 cooperates with Parkin for perinuclear clustering of depolarized mitochondria. Genes Cells 15:887–900

Pankiv S, Hoyvarde Clausen T, Lamark T, Brech A, Bruun JA, Outzen H, Øvervatn A, Bjørkøy G, Johansen T (2007) p62/SQSTM1 binds directly to Atg8/LC3 to facilitate degradation of ubiquitinated protein aggregates by autophagy. J Biol Chem 282:24131–24145

Poole AC, Thomas RE, Yu S, Vincow ES, Pallanck L (2010) The mitochondrial fusion-promoting factor mitofusin is a substrate of the PINK1/parkin pathway. PLoS One 5:e10054

Qu X, Yu J, Bhagat G, Furuya N, Hibshoosh H, Troxel A, Rosen J, Eskelinen E-L, Mizushima N, Ohsumi Y, Cattoretti G, Levine B (2003) Promotion of tumorigenesis by heterozygous disruption of the *beclin 1* autophagy gene. J Clin Invest 112:1809–1820

Rakovic A, Grunewald A, Seibler P, Ramirez A, Kock N, Orolicki S, Lohmann K, Klein C (2010) Effect of endogenous mutant and wild-type PINK1 on Parkin in fibroblasts from Parkinson disease patients. Hum Mol Genet 19:3124–3137

Riley BE, Kaiser SE, Shaler TA, Ng AC, Hara T, Hipp MS, Lage K, Xavier RJ, Ryu KY, Taguchi K, Yamamoto M, Tanaka K, Mizushima N, Komatsu M, Kopito RR (2010) Ubiquitin accumulation in autophagy-deficient mice is dependent on the Nrf2-mediated stress response pathway: a potential role for protein aggregation in autophagic substrate selection. J Cell Biol 191:537–552

Rubinsztein DC (2006) The roles of intracellular protein-degradation pathways in neurodegeneration. Nature 443:780–786

Seibenhener ML, Geetha T, Wooten MW (2007) Sequestosome 1/p62 more than just a scaffold. FEBS Lett 581:175–179

Shi G, Lee JR, Grimes DA, Racacho L, Ye D, Yang H, Ross OA, Farrer M, McQuibban GA, Bulman DE (2011) Functional alteration of PARL contributes to mitochondrial dysregulation in Parkinson's disease. Hum Mol Genet 20:1966–1974

Suzuki K, Ohsumi Y (2007) Molecular machinery of autophagosome formation in yeast, *Saccharomyces cerevisiae*. FEBS Lett 581:2156–2161

Suzuki K, Kubota Y, Sekito T, Ohsumi Y (2007) Hierarchy of Atg proteins in pre-autophagosomal structure organization. Genes Cells 12:209–218

Takahashi Y, Coppola D, Matsushita N, Cualing HD, Sun M, Sato Y, Liang C, Jung JU, Cheng JQ, Mul JJ, Pledger WJ, Wang HG (2007) Bif-1 interacts with Beclin 1 through UVRAG and regulates autophagy and tumorigenesis. Nat Cell Biol 9:1142–1151

Takamura A, Komatsu M, Hara T, Sakamoto A, Kishi C, Waguri S, Eishi Y, Hino O, Tanaka K, Mizushima N (2011) Autophagy-deficient mice develop multiple liver tumors. Genes Dev 25:795–800

Tanaka A, Cleland MM, Xu S, Narendra DP, Suen DF, Karbowski M, Youle RJ (2010) Proteasome and p97 mediate mitophagy and degradation of mitofusins induced by Parkin. J Cell Biol 191:1367–1380

Thoresen SB, Pedersen NM, Liestol K, Stenmark H (2010) A phosphatidylinositol 3-kinase class III sub-complex containing VPS15, VPS34, Beclin 1, UVRAG and BIF-1 regulates cytokinesis and degradative endocytic traffic. Exp Cell Res 316:3368–3378

Tsukada M, Ohsumi Y (1993) Isolation and characterization of autophagy-defective mutants of *Saccharomyces cerevisiae*. FEBS Lett 333:169–174

Vives-Bauza C, Zhou C, Huang Y, Cui M, de Vries RL, Kim J, May J, Tocilescu MA, Liu W, Ko HS, Magrane J, Moore DJ, Dawson VL, Grailhe R, Dawson TM, Li C, Tieu K, Przedborski S (2010) PINK1-dependent recruitment of Parkin to mitochondria in mitophagy. Proc Natl Acad Sci USA 107:378–383

Wang H, Song P, Du L, Tian W, Yue W, Liu M, Li D, Wang B, Zhu Y, Cao C, Zhou J, Chen Q (2011) Parkin ubiquitinates Drp1 for proteasome-dependent degradation: implication of dysregulated mitochondrial dynamics in Parkinson's disease. J Biol Chem 286:11649–11658

White E, Karp C, Strohecker AM, Guo Y, Mathew R (2010) Role of autophagy in suppression of inflammation and cancer. Curr Opin Cell Biol 22:212–217

Wooten MW, Hu X, Babu JR, Seibenhener ML, Geetha T, Paine MG, Wooten MC (2006) Signaling, polyubiquitination, trafficking, and inclusions: Sequestosome 1/p62's role in neurodegenerative disease. J Biomed Biotechnol 2006:1–12

Xie Z, Klionsky DJ (2007) Autophagosome formation: core machinery and adaptations. Nat Cell Biol 9:1102–1109

Yoshii SR, Kishi C, Ishihara N, Mizushima N (2011) Parkin mediates proteasome-dependent protein degradation and rupture of the outer mitochondrial membrane. J Biol Chem 286:19630–19640

Yoshikawa Y, Ogawa M, Hain T, Yoshida M, Fukumatsu M, Kim M, Mimuro H, Nakagawa I, Yanagawa T, Ishii T, Kakizuka A, Sztul E, Chakraborty T, Sasakawa C (2009) *Listeria monocytogenes* ActA-mediated escape from autophagic recognition. Nat Cell Biol 11:1233–1240

Youle RJ, Narendra DP (2011) Mechanisms of mitophagy. Nat Rev Mol Cell Biol 12:9–14

Yue Z, Jin S, Yang C, Levine AJ, Heintz N (2003) Beclin 1, an autophagy gene essential for early embryonic development, is a haploinsufficient tumor suppressor. Proc Natl Acad Sci USA 100:15077–15082

Zhu J, Chu CT (2010) Mitochondrial dysfunction in Parkinson's disease. J Alzheimers Dis 20 (Suppl 2):S325–S334

Ziviani E, Tao RN, Whitworth AJ (2010) Drosophila parkin requires PINK1 for mitochondrial translocation and ubiquitinates mitofusin. Proc Natl Acad Sci USA 107:5018–5023

The role of the co-chaperone BAG3 in selective macroautophagy: implications for aging and disease

Christian Behl

Abstract Maintenance of protein homeostasis, correct protein folding, refolding and clearance is of central importance for the function and survival of every cell. Here, the degradation of proteins is of particular importance, especially during aging and certain degenerative disorders when the protein load is increased. During cellular aging as well as under acute stress, there is a reciprocal change in expression of two members of the BAG (Bcl-2-associated athanogene) family, BAG1 and BAG3. While BAG1 serves an important function during the degradation of ubiquitinated proteins via the proteasome, BAG3 is the mediator of a novel macroautophagy pathway. This BAG3-mediated macroautophagy is based on the specificity of heat shock protein (HSP) 70 for misfolded proteins and also involves other protein partners, such as HSPB8, sequestosome-1/p62 (SQSTM1/p62) and the autophagosome protein LC3. BAG3 directly mediates the targeting and transport of degradation-prone substrates into aggresomes via the microtubule-motor dynein and also works independently of substrate ubiquitination.

Introduction

The proteins of a cell are in constant turnover. They are synthesized, properly folded, targeted to specific intracellular sites where needed, and have a certain half-life before they are degraded. This physiological life cycle is maintained and is described as protein homeostasis; it is crucial for the function and survival of any particular cell, ranging from highly proliferating to differentiated cells. Throughout their lifetimes, cells face permanently changing and, under certain circumstances, rather unfriendly conditions that also challenge the protein turnover system. As a

C. Behl (✉)
Institute for Pathobiochemistry, University Medical Center, Johannes Gutenberg University, Mainz 55099, Germany
e-mail: cbehl@uni-mainz.de

R.I. Morimoto and Y. Christen (eds.), *Protein Quality Control in Neurodegenerative Diseases*, Research and Perspectives in Alzheimer's Disease,
DOI 10.1007/978-3-642-27928-7_7,

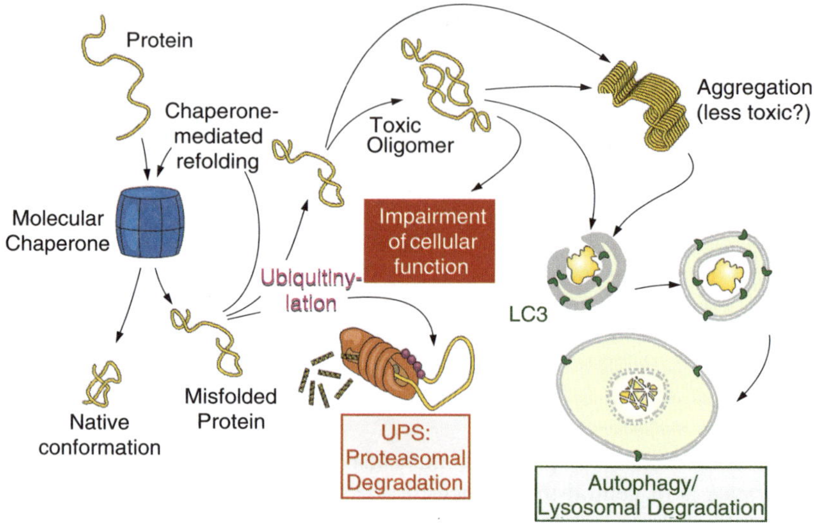

Fig. 1 The two principal pathways of protein degradation: ubiquitin-proteasome system and (macro)autophagy

consequence, protein degradation activity is especially under pressure and cells invest quite a lot of energy and effort to provide efficient protein degradation. In fact, cells have basically two separate protein degradation systems that differ, for instance, in their mode of action, degradation efficacy and substrate specificity. These two systems are the ubiquitin-proteasome system (UPS) and autophagy (here also referred to as macroautophagy; Nandi et al. 2006; Mizushima 2007; Fig. 1). While the proteasome system is rather well understood and defined, the macroautophagy pathway consists of a variety of separate degradation routes, involving a set of chaperones, co-chaperones, and adaptor proteins. In particular, the role of co-chaperones in macroautophagy has recently gained a lot of attention.

Increasing demand for protein degradation during aging and disease

Enhanced protein aggregation is a prominent hallmark of aged cells and is associated and partially linked to the onset and progression of neurodegenerative disorders (Rautou et al. 2010; Dillin and Cohen 2011). Protein aggregates are found in age-associated neurodegenerative disorders, including Huntington's disease, Parkinson's disease, and amyotrophic lateral sclerosis (ALS). Moreover, aggregated proteins are detected as pathological hallmarks in different non-neuronal diseases. For instance, protein deposits are found in various liver pathologies, known as Mallory-Denk bodies. Obviously, under the unfavorable

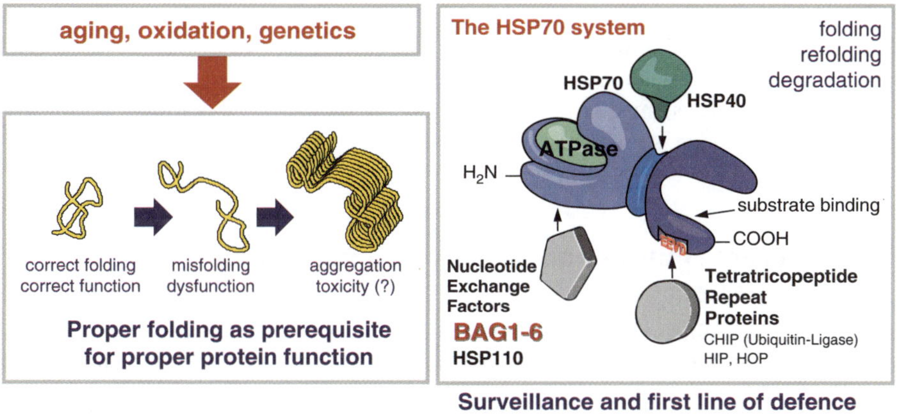

Fig. 2 The HSP70 system, a first line of defense: HSP70 controls proper folding and refolding of proteins, the prerequisite of correct protein function. During aging, proteins are substrates of oxidation, but genetic mutations can also lead to protein misfolding. Misfolding is recognized by the HSP70 system, which acts in concert with co-factors

conditions occurring during aging and in disease, cells are no longer able to fully degrade accumulating proteins and deposit this protein junk inside the cell. Oxidative modifications of all cellular biomolecules are observed during aging (Stadtman 2006). One view is that the intracellular deposition of protein aggregates is a consequence of a wear-and-tear process driven by, for instance, oxidations that can be seen in aged cells as deposits of lipufuscin. During disease, genetic and biochemical changes (e.g., mutations, folding errors) may also lead to the increased propensity of proteins to aggregate. The HSP70 system of chaperones and co-chaperones appears to be involved in protein degradation, besides being responsible for folding and refolding of proteins (Kampinga and Craig 2010; Young 2010; Fig. 2). In the neurodegenerative disease ALS, mutated forms of the antioxidant enzyme superoxide dismutase 1 (SOD1) have a tendency to aggregate. It has not yet been resolved whether the aggregation of mutated SOD1 is indeed the cause of neurotoxicity and, therefore, disease or a downstream event. In fact, the neurotoxic effects have also been described to be partially independent of the extent of aggregation (Witan et al. 2008, 2009). Nevertheless, deposition of protein junk might cause secondary effects and is at least an additional challenge for the physiology of the cell, rendering it potentially more vulnerable. A clear, visible sign of age-associated protein deposition is the occurrence of lipofuscin, a not-well defined collection of crosslinked proteins and lipids that can be seen in epidermal cells of the skin but is also found in other tissues (e.g., liver, brain). Interestingly, activation of the autophagy pathways with the drug rapamycin leads to enhanced clearance of aggregates of mutated huntingtin, the hallmark of Huntington's disease (Rose et al. 2010). The induction of autophagy pathways with small molecules is currently being discussed as one novel therapeutic option for neurodegenerative disorders accompanied by intracellular protein aggregation and deposition

(Fleming et al. 2011). But since different routes of intracellular autophagy are known and enhanced autophagy turnover has been linked to cancer development and progression (Levine and Kroemer 2008), a specific activation of selective autophagy pathways, in contrast to the general induction of various autophagy pathways all at once, should be considered.

The different routes of autophagy in a nutshell

In general, autophagy is a ubiquitous and evolutionarily conserved process in eukaryotes that degrades cytosolic components by the lysosome (Johansen and Lamark 2011). So far three different types of autophagy have been described in detail: macroautophagy, microautophagy and chaperone-mediated autophagy (CMA; Mizushima 2007; Arias and Cuervo 2011). The autophagy process itself also includes a degradation pathway called mitophagy that selectively eliminates organelles (mitochondria) to regulate their number and maintain organelle homeostasis (Youle and Narendra 2011). Autophagy (referred to hereafter as macroautophagy) was initially thought to be an unspecific bulk and robust degradation process. Based on the identification of various molecular players and modifiers, it is now clear that macroautophagy pathways exist that remove substrates in a highly selective manner (Johansen and Lamark 2011). In fact, macroautophagy is not a crude digestion process but follows a sophisticated, multi-step mechanism and interplay of membranes and proteins. Cytosolic material, for instance misfolded and aggretated proteins, is sequestered in a double-layered membrane structure, the so-called autophagosome, and is delivered to the degradation compartment, the lysosome. While macroautophagy has been considered for quite some time to be a process lacking substrate specificity, microautophagy involves direct sequestration of cytosolic components by invagination of the lysosomal membrane. In contrast, CMA displays a high selectivity that removes only a distinct subset of proteins carrying a pentapeptide lysosome-targeting motif (KFERQ). Such substrates are directly translocated into the lysosome after binding to the lysosomal receptor LAMP2A. For the translocation into the lysosome, substrates are then unfolded by a chaperone complex containing HSC70 and the co-chaperones BAG1, Hip, Hop, and HSP40/DNAJB1 (Arias and Cuervo 2011). Focusing on these autophagy types, macroautophagy in particular has recently attracted a lot of attention because it is becoming increasingly clear that 1) in addition to the UPS, macroautophagy plays an essential role in maintaining protein homeostasis, 2) macroautophagy can also display a selectivity in substrate clearance, and 3) an interplay of chaperones and co-chaperones mediates the selection of degradation-prone proteins (Dice 2007; Carra et al. 2008a; Johansen and Lamark 2011; Behl 2011). Indeed, a huge step forward was the discovery of molecular adaptors of autophagy, the proteins SQSTM1/p62 and NBR1 (neighbor of Brc1). These adaptors play a central role as cargo receptors and functional interfaces for the degradation of ubiquitinated protein substrates.

Moreover, the finding of the direct interaction between SQSTM1/p62 and NBR1 with LC3, a protein of the autophagosome, has shed new light on the macroautophagy route (Johansen and Lamark 2011).

BAG1 and BAG3 are reciprocally expressed and display different functions

While investigating basic biochemical differences of young and old cells, we found the expression of two members of the BAG family, BAG1 and BAG3, to be reciprocally regulated during cellular aging as well as under acute stress (e.g., oxidative stress or proteasomal inhibition). This reciprocal expression of these two co-chaperones is called the "BAG1/BAG3-switch" (Gamerdinger et al. 2009; Behl 2011). Under normal physiological conditions, protein quality control (PQC) is mainly mediated by the activity of HSP70 towards misfolded proteins. In this process, BAG1 obviously plays a major role, since the interaction of BAG1 with HSP70 is essential for substrate transfer and proteasomal degradation of poly-ubiquitinated (polyUb)-proteins (Gamerdinger et al. 2009). When the cellular environment changes and becomes unfriendly, representing pathophysiological conditions such as during aging and acute stress, an accumulation of misfolded and aggregated proteins can be observed. Moreover, the proteasomal degradation efficacy is decreased during aging and under non-physiological conditions. Here, BAG3 expression is increased and, most importantly, as a consequence of the accumulation of degradation-prone proteins, the BAG3-mediated selective macroautophagy is turned on. This "on-demand macroautophagy" is potentially of vital importance for cell survival and is an adaptive response.

HSP70 provides substrate specificity

An essential prerequisite of BAG3-mediated macroautophagy is the cooperation with HSP70. In fact, substrate specificity for misfolded proteins is based on the particular function of HSP70. The HSP70 chaperone system is complex, consisting of the main chaperone HSP70 and a larger group of co-chaperones and co-regulators (Hartl and Hayer-Hartl 2002; Lanneau et al. 2010; Young 2010), and it represents the intracellular detection system for unfolded and misfolded substrates. It specifically detects such substrates and transfers them to the intracellular protein degradation machinery. This substrate identification occurs via the activity of molecular chaperones that have a high binding affinity for solvent-exposed, unstructured and hydrophobic regions of non-native proteins. Under unfavorable or protein-denaturing conditions, including heat stress, many molecular chaperones are up-regulated. Therefore, these proteins are referred to as stress proteins or heat-shock proteins (HSP; Hartl and Hayer-Hartl 2002; Lanneau et al. 2010; Young 2010). When HSP70 binds to substrates, unstructured polypeptides can fold (or refold) into the native and functional three-dimensional structure, but, when this native state cannot be reached, as in the case of disease-related mutated proteins (e.g., polyQ-expanded huntingtin, mutant SOD1, α-synuclein), molecular chaperones change their function and can promote subsequent substrate degradation (Hartl and Hayer-Hartl 2002; Chang et al. 2007; Tang et al. 2007). The binding of substrates to HSP70 and their subsequent release is regulated by ATP in an ATP-consuming cycle. When ATP is bound to HSP70, binding of substrate occurs rather rapidly. In this state, the hydrolysis of bound ATP to ADP stabilizes the direct interaction of substrate with HSP70. The following release of bound ADP from HSP70 and the subsequent rebinding of a novel molecule ATP induce the dissociation of the chaperone-substrate complex and the release of the substrate. The activity of HSP70 in substrate folding and refolding, as well as the degradation activity, is regulated by co-chaperones that influence the ATP-consuming cycle (e.g., HSP40, HIP). Moreover, other HSP70-binding co-factors play a role, including the ubiquitin ligases CHIP and parkin (Young 2010; Fig. 2).

BAG3-mediated selective macroautophagy

Proteins of the BAG (Bcl-2-associated athanogene) protein family, which includes in total six members in humans (Takayama and Reed 2001), are key mediators of HSP70-assisted protein degradation pathways. The evolutionarily conserved BAG domain allows these proteins to interact directly with HSP70 and modify and regulate HSP70 function. BAG proteins directly affect the ATP-consuming cycle and regulate the ATP/ADP exchange on HSP70. Thereby, they trigger the release of a chaperone-bound substrate from HSP70 (Sondermann et al. 2001). Subsequently, BAG proteins may couple the released substrates to specific downstream cellular processes, such as the degradation pathways. For instance, BAG1 couples HSP70 to the proteasomal degradation pathway and mediates the transfer of substrate to the proteasome. BAG3, on the other hand, has been recently shown to specifically control HSP70-assisted protein degradation by the main alternative protein degradation route, macroautophagy (Carra et al. 2008b; Gamerdinger et al. 2009). Although BAG3, while directly interacting with HSP70, is at the core of this newly identified selective macroautophagy pathway, additional players, such as other co-chaperones, are necessary. The recruitment of the BAG3-mediated macroautophagy pathway involves polyUb-substrates, HSP70, HSPB8, SQSTM1/p62 and LC3 (Carra et al. 2008a; Gamerdinger et al. 2009). In fact, the multi-chaperone protein complex BAG3-HSPB8-HSP70 was found to control selective degradation of substrates such as polyUb-proteins and disease proteins, including polyQ-huntingtin and superoxide dismutase-1 (SOD1) mutants linked to ALS (Crippa et al. 2010; Gamerdinger et al. 2011). Interestingly, more recently, we found that the substrates for this BAG3-driven macroautophagy pathway do not

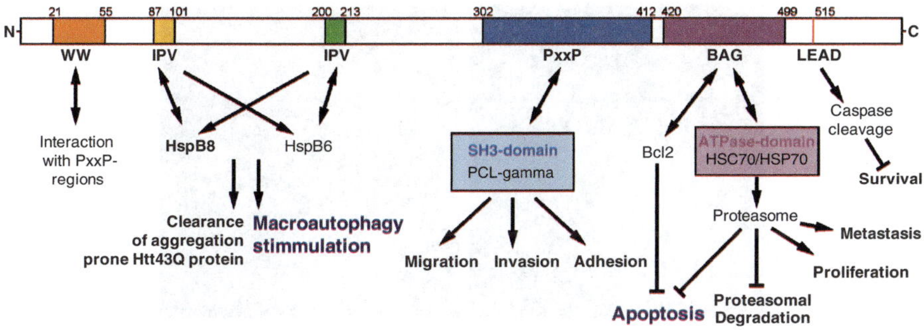

Fig. 3 BAG3 protein domains, interaction partners and functional links. Due to its domain structure, BAG3 is a highly promiscuous protein and is linked to key cellular processes, including apoptosis, proliferation and, as pointed out, to selective macroautophagy (Modified from McCollum et al. 2010)

necessarily have to be ubiquitinated, since non ubiquitinated proteins can also be clients of this particular BAG3-mediated macroautophagy pathway (Gamerdinger et al. 2011). Due to its domain structure, BAG3 can interact with many protein partners, linking it not only to autophagy but also to a variety of other key cellular pathways, including apoptosis and proliferation (McCollum et al. 2010; Fig. 3).

BAG3 controls aggresome targeting and interacts with dynein

It has been known for quite some time that damaged, misfolded and aggregated proteins (as autophagy substrates) are not randomly distributed throughout the cell but are found at specific intracellular sites. Indeed, selective macroautophagy strongly requires the concentration and separation of degradation substrates away from other cytosolic components. The sequestration of degradation substrates in a special perinuclear compartment, called the aggresome, is realized by a retrograde transport of the substrates along microtubules via the cytoplasmic dynein motor complex (Kopito 2000). Interestingly, we have found that the co-chaperone BAG3 directly associates with the microtubule motor dynein and mediates the selective transport of misfolded proteins to the aggresome. The binding of BAG3 to dynein occurs through a PxxP domain at BAG3 (Gamerdinger et al. 2011). Interestingly, the inhibition of dynein-mediated transport and, consequently, the prevention of aggresome formation lead to an inefficient degradation of aggregation-prone proteins by macroautophagy and are associated with the progression of various disorders, including ALS (Ravikumar et al. 2005). BAG3 is directly colocalized with aggresome markers (Fig. 4). An enhanced BAG3 expression increases the aggresome formation, whereas its knockdown by RNAi reduces the number of perinuclear aggresomes. Therefore, the following model is proposed: BAG3

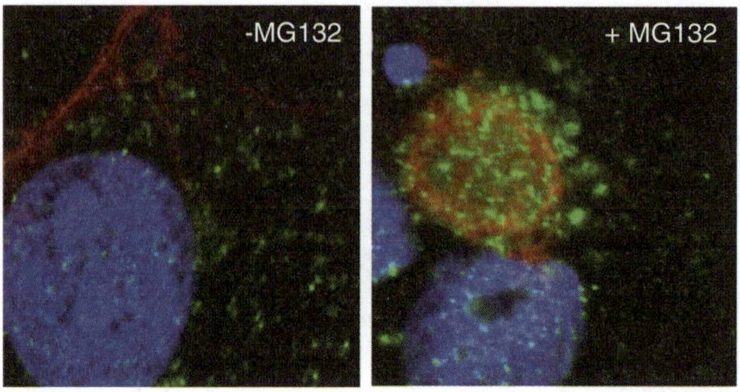

Fig. 4 BAG3 colocalizes with perinuclear aggresomes. Following impairment of the ubiquitin-proteasome system and accumulation of poly-ubiquitinated degradation substrates, the formation of perinculear aggresomes can be observed. Moreover, the expression of BAG3 is increased (*right*) and BAG3 (*green staining*) is colocalized with the aggresome compartment, depicted here via a vimentin staining (*red*). The green BAG3 staining is observed throughout the cell when no proteasome inhibitor (MG132) is added (*left*) and concentrated at the perinuclear site when MG132 is added

functions as an ATP/ADP-exchange factor and, thus, directly stimulates the release and transfer of substrate from HSP70, which provides substrate selectivity to the dynein motor complex. Subsequently, the degradation substrates are transported along the dynein/microtubules complex to a perinuclear site. Therefore, BAG3 directly promotes the shuttle of misfolded proteins to the aggresome and their concentration at a perinuclear site. In a direct interaction with SQSTM1/p62 and later on with LC3, the transfer into the macroautophagy pathway is possible. A link of aggresomes to the macroautophagy pathway has been described (Johnston et al. 1998; Webb et al. 2004). BAG3 in complex with HSPB8 and HSP70 serves a key function in the sequestration of degradation substrates and their transport into aggresomes and may also serve as a direct link to SQSTM1/p62, which binds to ubiquitin as well as to LC3 to switch on macroautopahgy. This process is highly relevant for disorders that are associated with protein aggregrates. Disease proteins, such as polyQ-huntingtin or mutated SOD1, cannot be handled by the proteasome and therefore accumulate. Consequently, the BAG3-HSP8-HSP70 multi-chaperone complex promotes targeting of these protein aggregates to the aggresome and their subsequent degradation. As already mentioned, enhanced protein aggregation is a hallmark of aging and various neurodegenerative diseases. It is well known that targeting of substrates to the aggresome is achieved by the ubiquitination of degradation-prone proteins that are recognized by ubiquitin binding proteins, such as the deacetylase HDAC6, binding to dynein (Kawaguchi et al. 2003; Yao 2010; Li et al. 2011). Because many misfolded proteins are not ubiquitinated, selective loading of cargo onto dynein must also occur independently of ubiquitination. Here, the co-chaperone BAG3 might play the key role.

In summary, the described novel pathway of aggresome targeting and selective macroautophagy mediated by BAG3 in concert with other co-chaperones is an important example of the involvement of chaperone molecules in autophagy and of a fine-tuned mechanism of macroautophagy. Selective macroautophagy mediated by specific co-chaperones could be considered a novel target for prevention and therapy of age-associated neurodegeneration in the future.

References

Arias E, Cuervo AM (2011) Chaperone-mediated autophagy in protein quality control. Curr Opin Cell Biol 23:184–189

Behl C (2011) BAG3 and friends: co-chaperones in selective autophagy during aging and disease. Autophagy 7:795–798

Carra S, Seguin SJ, Landry J (2008a) HspB8 and Bag3: a new chaperone complex targeting misfolded proteins to macroautophagy. Autophagy 4:237–239

Carra S, Seguin SJ, Lambert H, Landry J (2008b) HspB8 chaperone activity toward poly (Q)-containing proteins depends on its association with Bag3, a stimulator of macroautophagy. J Biol Chem 283:1437–1444

Chang HC, Tang YC, Hayer-Hartl M, Hartl FU (2007) SnapShot: molecular chaperones, Part I. Cell 128:212

Crippa V, Sau D, Rusmini P, Boncoraglio A, Onesto E, Bolzoni E, Galbiati M, Fontana E, Marino M, Carra S, Bendotti C, De Biasi S, Poletti A (2010) The small heat shock protein B8 (HspB8) promotes autophagic removal of misfolded proteins involved in amyotrophic lateral sclerosis (ALS). Hum Mol Genet 19:3440–3456

Dice JF (2007) Chaperone-mediated autophagy. Autophagy 3:295–299

Dillin A, Cohen E (2011) Ageing and protein aggregation-mediated disorders: from invertebrates to mammals. Philos Trans R Soc Lond B Biol Sci 366:94–98

Fleming A, Noda T, Yoshimori T, Rubinsztein DC (2011) Chemical modulators of autophagy as biological probes and potential therapeutics. Nat Chem Biol 7:9–17

Gamerdinger M, Hajieva P, Kaya AM, Wolfrum U, Hartl FU, Behl C (2009) Protein quality control during aging involves recruitment of the macroautophagy pathway by BAG3. EMBO J 28:889–901

Gamerdinger M, Kaya AM, Wolfrum U, Clement AM, Behl C (2011) BAG3 mediates chaperone-based aggresome-targeting and selective autophagy of misfolded proteins. EMBO Rep 12:149–156

Hartl FU, Hayer-Hartl M (2002) Molecular chaperones in the cytosol: from nascent chain to folded protein. Science 295:1852–1858

Johansen T, Lamark T (2011) Selective autophagy mediated by autophagic adapter proteins. Autophagy 7:279–296

Johnston JA, Ward CL, Kopito RR (1998) Aggresomes: a cellular response to misfolded proteins. J Cell Biol 143:1883–1898

Kampinga HH, Craig EA (2010) The HSP70 chaperone machinery: J proteins as drivers of functional specificity. Nat Rev Mol Cell Biol 11:579–592

Kawaguchi Y, Kovacs JJ, McLaurin A, Vance JM, Ito A, Yao TP (2003) The deacetylase HDAC6 regulates aggresome formation and cell viability in response to misfolded protein stress. Cell 115:727–738

Kopito RR (2000) Aggresomes, inclusion bodies and protein aggregation. Trends Cell Biol 10:524–530

Lanneau D, Wettstein G, Bonniaud P, Garrido C (2010) Heat shock proteins: cell protection through protein triage. Scientific World Journal 10:1543–1552

Levine B, Kroemer G (2008) Autophagy in the pathogenesis of disease. Cell 132:27–42

Li G, Jiang H, Chang M, Xie H, Hu L (2011) HDAC6 alpha-tubulin deacetylase: a potential therapeutic target in neurodegenerative diseases. J Neurol Sci 304:1–8

McCollum AK, Casagrande G, Kohn EC (2010) Caught in the middle: the role of Bag3 in disease. Biochem J 425:e1–e3
Mizushima N (2007) Collaboration of proteolytic systems. Autophagy 3:179–180
Nandi D, Tahiliani P, Kumar A, Chandu D (2006) The ubiquitin-proteasome system. J Biosci 31:137–155
Rautou PE, Mansouri A, Lebrec D, Durand F, Valla D, Moreau R (2010) Autophagy in liver diseases. J Hepatol 53:1123–1134
Ravikumar B, Acevedo-Arozena A, Imarisio S, Berger Z, Vacher C, O'Kane CJ, Brown SD, Rubinsztein DC (2005) Dynein mutations impair autophagic clearance of aggregate-prone proteins. Nat Genet 37:771–776
Rose C, Menzies FM, Renna M, Acevedo-Arozena A, Corrochano S, Sadiq O, Brown SD, Rubinsztein DC (2010) Rilmenidine attenuates toxicity of polyglutamine expansions in a mouse model of Huntington's disease. Hum Mol Genet 19:2144–2153
Sondermann H, Scheufler C, Schneider C, Hohfeld J, Hartl FU, Moarefi I (2001) Structure of a Bag/Hsc70 complex: convergent functional evolution of Hsp70 nucleotide exchange factors. Science 291:1553–1557
Stadtman ER (2006) Protein oxidation and aging. Free Radic Res 40:1250–1258
Takayama S, Reed JC (2001) Molecular chaperone targeting and regulation by BAG family proteins. Nat Cell Biol 3:E237–E241
Tang YC, Chang HC, Hayer-Hartl M, Hartl FU (2007) SnapShot: molecular chaperones, Part II. Cell 128:412
Webb JL, Ravikumar B, Rubinsztein DC (2004) Microtubule disruption inhibits autophagosome-lysosome fusion: implications for studying the roles of aggresomes in polyglutamine diseases. Int J Biochem Cell Biol 36:2541–2550
Witan H, Kern A, Koziollek-Drechsler I, Wade R, Behl C, Clement AM (2008) Heterodimer formation of wild-type and amyotrophic lateral sclerosis-causing mutant Cu/Zn-superoxide dismutase induces toxicity independent of protein aggregation. Hum Mol Genet 17:1373–1385
Witan H, Gorlovoy P, Kaya AM, Koziollek-Drechsler I, Neumann H, Behl C, Clement AM (2009) Wild-type Cu/Zn superoxide dismutase (SOD1) does not facilitate, but impedes the formation of protein aggregates of amyotrophic lateral sclerosis causing mutant SOD1. Neurobiol Dis 36:331–342
Yao TP (2010) The role of ubiquitin in autophagy-dependent protein aggregate processing. Genes Cancer 1:779–786
Youle RJ, Narendra DP (2011) Mechanisms of mitophagy. Nat Rev Mol Cell Biol 12:9–14
Young JC (2010) Mechanisms of the Hsp70 chaperone system. Biochem Cell Biol 88:291–300

Predicting fates in models of neurodegenerative disease: longitudinal measures of protein homeostasis in live neurons

Steven Finkbeiner

Abstract Abnormal aggregation and deposition of proteins may be common features of neurodegenerative diseases. A prevailing hypothesis is that misfolded proteins stress the capacity of the protein homeostasis system and result in neurodegeneration. However, protein dyshomeostasis has been difficult to investigate by conventional approaches. To overcome these obstacles, we developed a new strategy that combines robotic microscopy and longitudinal analysis to quantitatively measure the relationship between intermediate disease-associated changes in a neuron and its ultimate fate at a single-cell level. Importantly, the method distinguishes between pathogenic, incidental, and coping mechanisms, and any particular mechanism, positive or negative, can be assessed for its prognostic power. With this method, we can construct predictive quantitative and temporally dynamic multivariate models of neurodegeneration that are useful for developing rational therapeutic strategies. Surprisingly, we found that neurons that form protein deposits called inclusion bodies (IBs) survive longer than those without IBs. Thus, IBs seem to be a coping mechanism. Recently, we used photoswitchable proteins and automated microscopy to measure protein clearance in live neurons. By measuring huntingtin (htt) metabolism in single neurons and relating it to each neuron's fate, we found that the stress experienced by a neuron's protein homeostasis system in response to mutant huntingtin (mhtt) predicted neurodegeneration better than the dose of mhtt it received. Our findings suggest a simple unifying model: polyglutamine expansions cause neurodegeneration by adopting conformations that stress the capacity of the protein homeostasis system, and strategies to promote protein homeostasis may be broadly neuroprotective.

S. Finkbeiner (✉)
Gladstone Institute of Neurological Disease, Taube-Koret Center for Huntington's Disease Research, Hellman Family Foundation Alzheimer's Disease Research Program, and the Departments of Neurology and Physiology, University of California, 1650 Owens St., Office 308, San Francisco, CA 94158, USA
e-mail: sfinkbeiner@gladstone.ucsf.edu

R.I. Morimoto and Y. Christen (eds.), *Protein Quality Control in Neurodegenerative Diseases*, Research and Perspectives in Alzheimer's Disease,
DOI 10.1007/978-3-642-27928-7_8,

Introduction

The field of neurodegenerative disease research owes pathologists a great debt. As clinicians attempted to define and distinguish neurological diseases, pathologists helped enormously by characterizing the histological and cytological abnormalities that correlated with specific clinical phenotypes (Jagadha and Becker 1989; Mackenzie et al. 2006; Pamphlett et al. 2009; Robitaille et al. 1997; Scheff and Price 2003). In some cases, this work led to the surprising discovery that patients who appeared to have nearly the identical disease based on its clinical features could be convincingly distinguished by the pathological features of their brain tissue. Soon pathological analysis became the gold standard for the diagnosis of conditions that carried a presumptive clinical diagnosis.

The dominant role of neuropathology in the definition of neurodegenerative disease, however, has unwittingly shaped the conceptualization of neurodegenerative diseases in ways that have not necessarily been entirely helpful. From the perspective of neuropathology, diseases came to be defined by the unfolding appearance of abnormalities in different brain regions over time. Evolving pathology became conflated with pathogenesis, and little consideration was given to the possibility that at least some of the abnormalities detected by pathologists might actually be helpful responses by the nervous system to the insults that triggered the disease process (Finkbeiner et al. 2006). The distinction is critical, too, because the stakes are so high. The series of recent high-profile clinical trial failures of drugs designed to treat Alzheimer's disease (AD) by targeting amyloid formation has raised the key question of whether amyloid is truly the critical drug target for AD (Maccioni et al. 2010; Mura et al. 2010; Robakis 2010; Struble et al. 2010). Could billions have been allocated more productively if our understanding of AD pathogenesis had exceeded our understanding of AD pathology?

Since neuropathologists were often called upon to make definitive diagnoses, they paid particular attention to features of diseases that distinguished one from another rather than to the substantial number of common threads that exist among neurodegenerative diseases. In turn, this view led to an emphasis on distinctions and even the perception that the major neurodegenerative diseases are fairly unrelated. Flowing from this paradigm, the focus of drug companies and funding agencies was on specific diseases rather than the biological mechanisms that might connect them. The common mechanisms that tie neurodegenerative diseases together may have received less attention than they deserved and opportunities for therapeutic intervention may have been missed (Beal et al. 2006).

Protein deposits

Our interest in neurodegenerative diseases began with what we thought was a simple aim. We wanted to resolve the pathogenic significance of a pathological hallmark of Huntington's disease (HD): inclusion body (IB) formation (Floyd and Hamilton 1999;

Sisodia 1998). Pathologists have described abnormal protein deposits in most major neurodegenerative diseases, including extracellular amyloid plaques and intracellular neurofibrillary tangles in AD, intraneuronal protein deposits called Lewy bodies in Parkinson's disease (PD) and Bunina bodies in amyotrophic lateral sclerosis (ALS), and intraneuronal IBs in frontotemporal dementia (FTD), Creutzfeld-Jacob disease and HD (Fink 1998; Goedert et al. 1998). From these studies, the pathological evidence for a role in pathogenesis seemed compelling (Alves-Rodrigues et al. 1998; Davies et al. 1997). With the exception of amyloid deposits, most of these pathological abnormalities were only seen in the context of bona fide disease, and the burden of these deposits generally correlated with the clinical severity of disease (Becher et al. 1998). These gross correlations led to the prevailing view that these deposits were the key cause of neurodegenerative diseases and represented a common mechanism that encompassed all the major neurodegenerative diseases.

In the late 1990s, increasing examples from preclinical models of these diseases began to cast doubt on this view. Observations or even experimental perturbations were reported in which neuronal death or the behavioral or pathological sequellae of neurodegeneration could be dissociated from IB formation (Klement et al. 1998; Saudou et al. 1998). For some, the demonstration experimentally that IB formation was not required for neurodegeneration was enough to shift their focus to other potential pathogenic mechanisms. However, skeptics pointed out that the perturbations themselves may have had unintended effects that changed the basic biology of pathogenesis and undermined the relevance of those results to the mechanisms of the human disease.

The skeptics had a point. Given the slow, stochastic, and cell-specific nature of most neurodegenerative diseases, how can we know with certainty that the abnormalities we detect and grossly correlate to disease contribute directly to pathogenesis? Could they also be incidental to disease or represent an adaptive response of the organism to the inciting insult?

After considerable reflection, we concluded that the skeptics were right. There was no way that a conventional "snap shot" of different cells at different time points during the course of a disease could resolve the pathogenic significance of abnormalities. The gross features of a coping response might closely mirror those of a pathogenic one — appearing only in the context of disease and increasing in proportion to the burden of disease (Finkbeiner et al. 2006). We decided that the only way to resolve these issues was to invent a new technology that permitted us to monitor the neurodegenerative disease process as it unfolded in live cells over time and then to measure the exact relationship between the appearance of abnormalities and a fate of interest.

Now in its third generation, our new technology is known as a robotic microscope (Arrasate and Finkbeiner 2005; Daub et al. 2009). Fundamentally, the system was designed to enable the investigator to model neurodegenerative disease as it unfolds in live neurons from rodents or differentiated from induced pluripotent stem cells. We accomplished this feat by attaching fluorescent proteins to disease-causing proteins. We then observed the levels of the construct and IB formation in real-time and measured quantitatively the relationship between those parameters

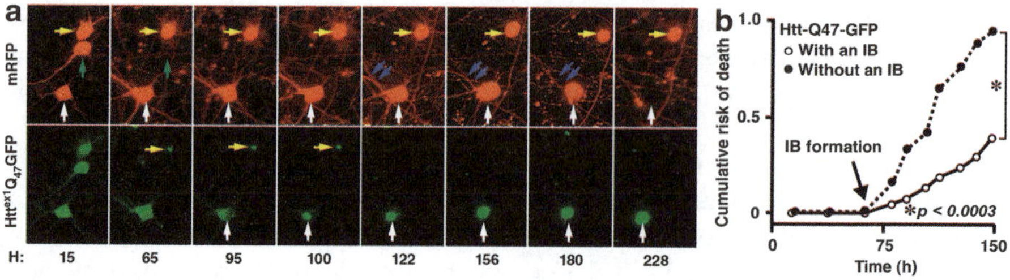

Fig. 1 (**a**) PolyQ expansion–dependent cell death measured with an automated microscope. Longitudinal tracking of single neurons expressing mRFP (*top panel*) and htt^{ex1}-Q47-GFP (*bottom panel*). Two neurons (*yellow and white arrows, top row*) that formed IBs (*yellow and white arrows, bottom row*) outlived a third neuron that died without an IB (*green arrow*). Soon after an IB formed (*white arrow, bottom panel*), mutant htt disappeared elsewhere in the neuron. Neuron morphology remained intact for days (*top row*), but then neurites degenerated (*blue arrows*) and the neuron died. (**b**) IB formation is associated with reduced death risk and increased survival among neurons transfected with htt^{ex1}-Q47-GFP that are alive beginning on the fourth day (n = 224 neurons, three experiments). (Adapted with permission from Arrasate et al. 2004)

and some sign of neuronal death or degeneration. Essential to its design, the software controlling the robotic microscope enabled the investigator to follow hundreds or even thousands of individual neurons over time and to relate cellular changes of putative pathogenic significance to a fate of unequivocal interest. Thankfully, the statistical tools, known as survival analysis, for quantifying such relationships were already well established for the fields of engineering and clinical medicine. They make it possible to measure with exquisite sensitivity the relationship between some intermediate change and an ultimate fate.

Applied to a faithful primary neuron model of HD, the technology provided a surprising result. IB formation was not associated with neurodegeneration caused by mhtt. In fact, cells that formed IBs had a lower risk of death; the neurons survived better (Arrasate et al. 2004; Miller et al. 2010; Fig. 1). It was as if IB formation was part of a coping mechanism rather than a pathogenic one. The result made sense in light of other findings from the same experiments: levels of forms of mhtt not bound in IBs (so-called diffuse mhtt) predicted whether and when specific striatal neurons died. As IBs formed, levels of diffuse mhtt dropped to near baseline levels. Thus, IB formation appeared to mitigate the toxicity of more diffuse forms of mhtt to the extent that it reduced their levels and sequestered them into an IB, which appeared relatively inert.

Though the result seemed clear-cut, it was less clear if the result had more widespread implications for even more relevant models of HD. To investigate this possibility, we developed human neuron models of HD by collecting fibroblasts from people harboring disease-associated HD mutations, reprogramming them with conventional methods to create induced pluripotent stem cells, and then applying to them a regimen of extracellular factors that caused them to differentiate into neurons and related cell types. The human neurons that resulted expressed full-length human htt at its endogenous level. Remarkably, we found that these cells showed a reduction in survival and abnormalities in neurite length that correlated to the severity of the HD-causing mutation and yet did not form any detectable IBs (unpublished observations). Thus, human neurons made from HD patients that expressed mhtt at physiological levels underwent degeneration without requiring IB formation.

IB formation is clearly not required for neurodegeneration in HD and appears to mitigate the process. What about other diseases? We developed primary neuron models of ALS/FTD based on familial mutations in Tar DNA binding protein 43 (TDP43; Barmada et al. 2010) and of PD based on synuclein or versions of leucine-rich repeat kinase 2 (LRRK2) harboring PD-causing mutations (Nakamura et al. 2011; Fig. 2). We showed that synuclein induced neurodegeneration without forming IBs and that, although TDP43 formed IBs, they were neither required nor did they predict neurodegeneration (Barmada et al. 2010; Nakamura et al. 2011; Fig. 2). In our model of LRRK2-dependent neurodegeneration, IBs formed but they were not required for neurodegeneration and, in some cases, appeared to lead to an improved outcome (unpublished observations). Minimally, these findings indicate that IBs are not the pathogenic species responsible for neurodegenerative disease. In some cases, they are demonstrably associated with better survival, suggesting that their formation may be a homeostatic response of neurons to the presence of proteins prone to misfold.

Toxic conformers

If IBs are not the pathogenic species, what is? The ability of the robotic microscope to generate longitudinal data from thousands of individual neurons allowed us to observe other reproducible changes that took place concurrently as IBs formed. The most striking discovery was that, almost coincident with the formation of an IB, levels of diffuse mhtt throughout the rest of the neuron fell to almost undetectable levels within about 1–2 days after the first sign of IB formation (Arrasate et al. 2004; Miller et al. 2010). Indeed, levels of diffuse mhtt within a neuron on the first day or two after the introduction of mhtt significantly predicted whether and when that neuron would die; the higher the starting levels, the shorter that cell would live. There was no such significant relationship for versions of htt without disease-causing mutations, suggesting that this dose–response relationship was specific to pathogenic mechanisms relevant to HD and that the toxic species of mhtt resides in the diffuse forms that get cleared when IBs form.

Specifying the nature of the toxic species has proved to be a tremendous technical challenge. Is it a misfolded monomer? An oligomeric aggregation intermediate? Is it even a "species" at all or perhaps one or more non-native protein folds that can be found to a varying extent in many conformations of mhtt? Does post-translational modification play a role? We and others have found residues within htt whose phosphorylation dramatically regulates its toxicity, but the mechanism remains an enigma (Humbert et al. 2002). The tools to explore the effects of the disease-causing mutation on structure and assembly states in many ways are even

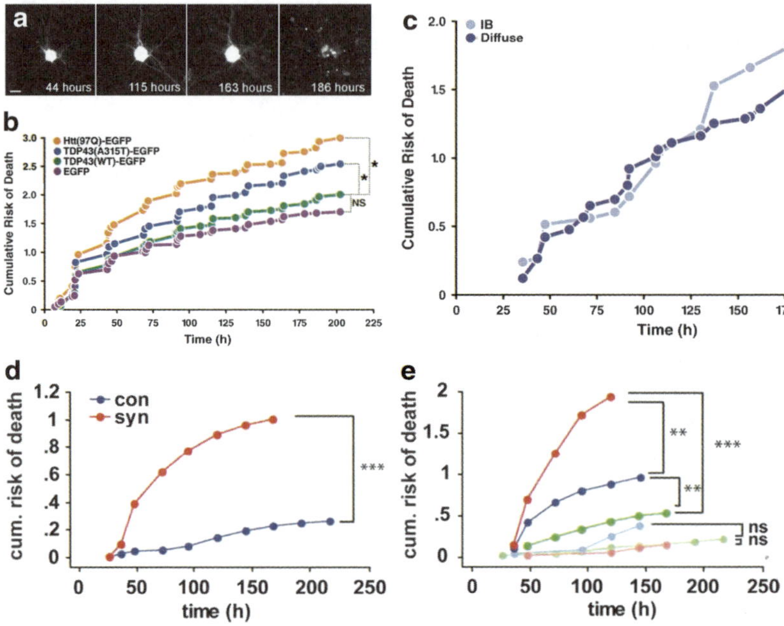

Fig. 2 Role of protein dyshomeostasis in primary neuron models of Amyotrophic Lateral Sclerosis and Parkinson's disease. (**a**) TDP43(A315T)-EGFP, but not TDP43(WT)-EGFP, is toxic in primary cortical neurons. Hundreds of individual neurons were consecutively imaged over an extended period. Neuronal death was marked by blebbing, rounding, or fragmentation of the cell body, as visualized by mCherry, a fluorescent co-transfection marker. (**b**) The time of death for each neuron was used to create a Kaplan-Meier survival curve, from which a cumulative hazard curve was constructed. *p < 0.0001 (log-rank test). NS, not significant. (**c**) Cumulative hazard curves of neurons with and without IBs expressing wild-type (WT) or mutant TDP43-EGFP. There is no significant difference in the risk of death between neurons with and without IBs expressing either WT or mutant TDP43-EGFP. *p < 0.01.**p < 0.001. NS, not significant. p values in C were determined by the log-rank test. (Adapted with permission from (Barmada et al. 2010)). (**d**) Rat hippocampal neurons were transfected 5 days after plating with mRFP and either synuclein (syn) or vector control (con). Cumulative risk of death curves indicate that primary neurons expressing synuclein have a significantly greater risk of death than control cells (***p < 0.0001, log-rank test). n = 112–115 neurons per group. The experiment was repeated three times with similar results. (**e**) Transfected neurons were stratified into three groups (*low, medium and high*) based on mRFP fluorescence as a surrogate for synuclein expression. a.u. = arbitrary units of fluorescence. Within each of these groups, synuclein (*dark*) and control (*light*) transfected neurons showed no significant difference in mRFP fluorescence. Cumulative risk of death curves indicate that neurons expressing higher levels of synuclein (*red and blue*) have a dose-dependent greater risk of death than controls expressing equivalent mRFP. **p < 0.005, ***p < 0.0001; ns, not significant, n = 22–46 neurons per group. (Adapted with permission from Nakamura et al. 2011)

more limited than those to study post-translational modifications (Ramdzan et al. 2010). Whereas purified mhtt can be coaxed to form myriad aggregation intermediates in a test tube, the tools are largely lacking to prove that any of these species exist in situ or to measure the biological activity of any one of these species and distinguish it from another.

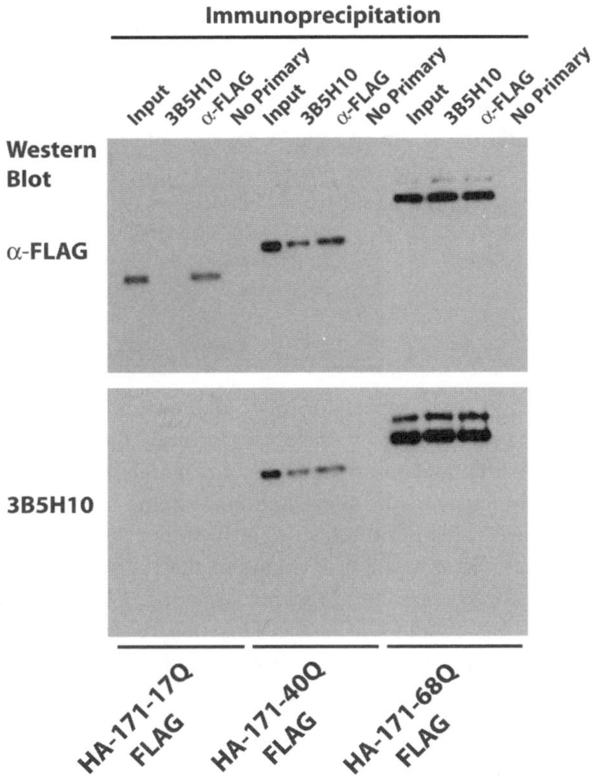

Fig. 3 Immunoprecipitation of huntingtin. Extracts from HEK 293 cells that contain HA-171-17-FLAG, HA-171-40-FLAG, or HA-171-68-FLAG were mixed together with anti-FLAG, 3B5H10, or no primary antibodies and then coprecipitated with protein G–Sepharose. Samples of the original extracts (Input) or the coprecipitated proteins were subjected to SDS-PAGE and blotted with either anti-FLAG or 3B5H10 antibodies as indicated. (Adapted with permission from Brooks et al. 2004)

To try to sort this out, we turned to immunology. We generated monoclonal antibodies against a 171 amino acid N-terminal fragment of mhtt and found six that recognized mutant better than wild-type htt (Brooks et al. 2004; Peters-Libeu et al. 2005; Fig. 3). We since demonstrated that one of these antibodies, 3B5H10, recognizes an epitope formed by the disease-associated polyglutamine expansion that causes HD when it is in a compact hairpin structure (Zhang et al. 2011). We used our robotic microscope to evaluate the prognostic significance of binding by 3B5H10 to mhtt in situ. Of all the conformation-specific antibodies, 3B5H10 was the only one that significantly predicted neurodegeneration and did so for versions of htt with polyglutamine stretches that spanned from those that are normal in length to those that cause severe disease (unpublished observations). Mutations positioned within the polyglutamine stretch in a way that would be predicted to disrupt β-hairpin formation reduced the toxicity of that version of htt and the ability of 3B5H10 to recognize it (Zhang et al. 2011). Remarkably, as mhtt aggregated into IBs, 3B5H10 lost its ability to bind mhtt, suggesting that the process of IB formation caused the hairpin epitope to be re-folded or buried. This observation also suggested a potential mechanism by which IB formation might mitigate mhtt-induced neurodegeneration. If the epitope or protein fold recognized by 3B5H10 is a toxic species or closely related to one, IB formation may mitigate toxicity by removing that epitope or protein fold from the cell.

Protein clearance

If removing or refolding toxic species of misfolded protein prevents degeneration, why don't neurons simply do it themselves through their network of protein chaperones or proteasomal and autophagic protein clearance pathways? The answer is that we don't know, but we speculate that the systems in neurons that refold or clear misfolded disease-causing proteins might be either comparatively insensitive to the problem or inadequate to the task. Many of these diseases are specific to the nervous system, despite the fact that, in some cases (e.g., HD), the disease-causing protein is expressed throughout the body. This observation led to the idea that neurons might regulate protein homeostasis differently than non-neuronal cells or might be particularly sensitive to misfolded proteins (Tsvetkov et al. 2009).

What happens to the ubiquitin proteasome system (UPS) in neurons during the process of neurodegeneration? Others reported that the UPS might be specifically affected by proteins prone to misfold (Alves-Rodrigues et al. 1998; Bedford et al. 2008; Floyd and Hamilton 1999; Mitra and Finkbeiner 2008). To investigate the dynamics of UPS function in neurodegenerative disease, we used a fluorescent reporter protein targeted to the proteasome and monitored its levels in our primary neuron model of HD with our automated microscope (Mitra et al. 2009). The fluorescent reporter is constitutively produced in the cell and degraded by the proteasome, so its steady state levels rise until production and clearance are balanced. Levels of the reporter increased from baseline steady-state levels in the days before IB formation, but once an IB formed, the levels of diffuse mhtt fell back to baseline (Fig. 4). There are at least two explanations for this phenomenon. First, diffuse mhtt might directly inhibit the proteasome, and IB formation might de-repress this inhibition by sequestering mhtt away from the proteasome and into IBs. Second, mhtt might increase the overall load of misfolded proteins in neurons. This increase could occur if mhtt used up critical capacity of the re-folding and clearance pathways, leaving the neuron less able to deal effectively with the load of protein misfolding that results from normal physiological processes. In this scenario, the accumulation of the reporter does not reflect a reduction in flux through the proteasome. Quite the opposite, flux through the proteasome may be functioning at maximum capacity, and accumulation would reflect competition for proteasomal clearance of the reporter by the increased load of other misfolded

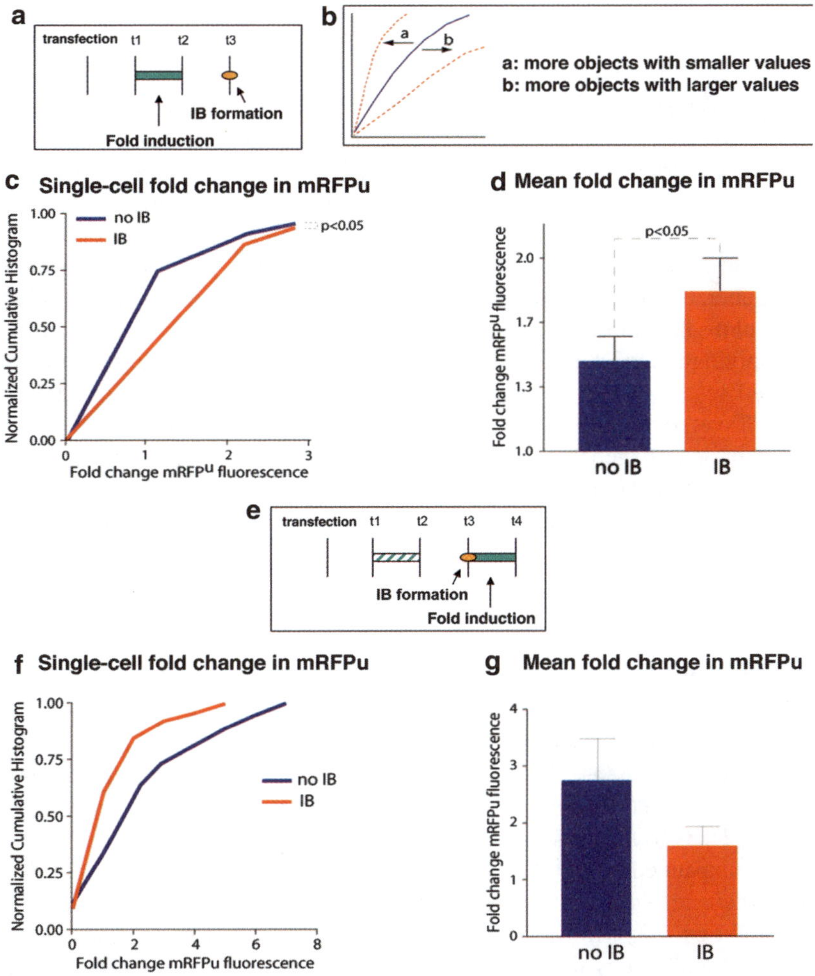

Fig. 4 Levels of a proteasome flux reporter (RFP^u) accumulate in neurons prior to, and then decline after, IB formation. Primary striatal neurons transfected with mhtt (Htt^{ex1}-Q72-GFP), a transfection and viability marker (BFP), and a reporter of proteasome flux consisting of a version of red fluorescent protein targeted to the proteasome by fusion to a degron (RFP^u). Individual neurons were followed longitudinally and analyzed for changes in (RFP^u) levels in relation to IB formation. (**a**) Cohorts of neurons that contained or lacked an IB were identified at a time point after transfection, and levels of RFP^u were measured in the two cohorts two days prior to IB formation. (**b–c**) The cumulative distributions of RFP^u levels were compared. (**c–d**) Neurons destined to form IBs showed significantly higher RFP^u levels prior to IB formation. (**e–g**) Using similar approaches, we followed levels of RFP^u after IBs formed and found that they fell relative to neurons without IBs. (Adapted with permission from Mitra et al. 2009)

proteins present in the neuron. We favor the latter interpretation. There is evidence that mhtt can be degraded by the proteasome, and proteasome inhibitors increase mhtt levels and promote rapid IB formation (Mitra et al. 2009).

These findings allow us to make some intriguing predictions. For example, pharmacological methods that de-repress cellular chaperone and protein clearance pathways might have therapeutic potential. Autophagy is an intriguing protein clearance pathway: it is constitutively active and can, at least in principle, clear protein aggregates that might harbor especially toxic conformations that stress the chaperone network (Shacka et al. 2008). Autophagy also is the only known pathway for clearing senescent or damaged organelles, such as mitochondria (Dagda et al. 2008; Kanki et al. 2009); the accumulation of damaged or dysfunctional mitochondria has emerged as another potential common mechanism mediating neurodegeneration (Beal et al. 2006; Beal 2005). We searched for small molecules that stimulate flux through the autophagy pathway in primary neurons. Interestingly, applying the most effective autophagy inducers to our primary neuron model of HD led to a reduction in levels of diffuse mhtt, IB formation and improved survival (Tsvetkov et al. 2010; Fig. 5). So whether or not autophagy is defective in HD or other neurodegenerative diseases, modest stimulation of this pathway might improve the clearance of misfolded proteins and improve the balance between the load of protein misfolding and the capacity of the cellular systems to deal with it.

A unifying model

We have synthesized these findings into a dynamic and predictive model of neurodegenerative disease that focuses on stress on the intrinsic protein homeostasis system (Balch et al. 2008; Powers et al. 2009; Fig. 6). In our model, protein misfolding leads to the presentation of epitopes or folds (e.g., the hairpin structure formed by expanded polyglutamine stretches) that require the binding of members of the protein refolding/chaperone network to prevent aggregation. These "toxic folds" might be found in misfolded monomeric versions of some disease-causing proteins or stabilized by the aggregation process and represented in specific aggregation intermediates (e.g., certain protein oligomers). The reduction in capacity of the refolding network that results leaves other metastable proteins in the cell prone to misfolding. Widespread protein misfolding and the pleiotropic loss of functions of affected proteins could lead to the neuronal dysfunction and death that characterizes neurodegenerative disease. In that context, IB formation could help neurons cope with protein misfolding by sequestering forms of the protein that cause stress on the re-folding network into IBs where the "toxic folds" are no longer present or available. In effect, IB formation might be a rapid way to create spare capacity to render misfolded proteins relatively inert. It might give a neuron time to upregulate proteasomal or autophagic protein clearance pathways.

Viewed from this perspective, IBs represent a "common thread" of neurodegenerative disease — not a common pathogenic mechanism, but rather a common coping mechanism by neurons to deal with misfolded proteins that exceed the capacity of the protein homeostasis system to refold or clear them.

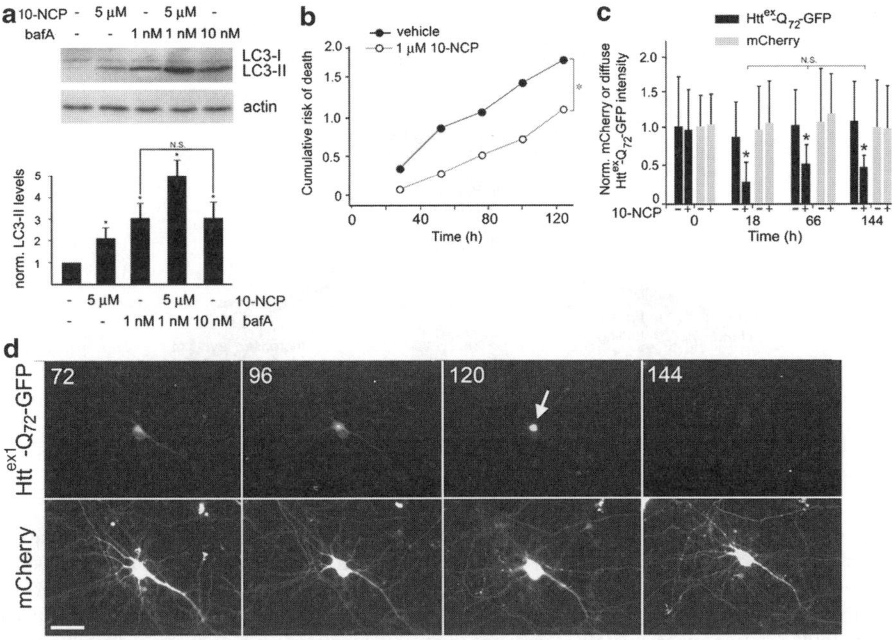

Fig. 5 N10-substituted phenoxazine (10-NCP) induces autophagy in primary neurons and protects them from mhtt-induced neurodegeneration. (**a**) Autophagy was efficiently induced in striatal neurons by 10 μM 10-NCP, as reflected by the increased levels of LC3-II. Actin was used as a loading control throughout this figure. *p < 0.001 (pairwise *t* test) (see note below). norm., normalized. LC3-II accumulation in striatal neurons treated with 5 μM 10-NCP with or without 1 nM bafilomycin A [10-NCP, overnight; bafilomycin A (bafA) was then added for 4 h]. Bafilomycin A reached a ceiling effect, and higher concentrations (e.g., 10 nM) did not further increase LC3-II levels. LC3-II increased in 10-NCP-treated cells when bafilomycin A was added. These findings suggest that 10-NCP is inducing autopaghy and not simply causing accumulation of LC3-II by inhibiting its lysosomal clearance. *p < 0.001 (ANOVA). (**b**) Striatal neurons transfected with mCherry and mHtt^{ex1}-GFP were treated with 1 μM 10-NCP or vehicle. Cumulative risk of death was calculated from Kaplan–Meier curves. 10-NCP reduced the risk for death (i.e., improved survival) of neurons expressing mutant Httex1. *p <0.001 (Mantel–Cox test). (**c**) Levels of diffuse mutant Htt and mCherry from neurons transfected and treated as in **b**. After exclusion of neurons with IBs, GFP or mCherry levels were normalized (Norm.) to fluorescence at the first time point. The levels of cotransfected mCherry did not change.*p < 0.01 (ANOVA). N.S., not significant. (**d**) IB (arrow) that spontaneously disappeared during the experiments in C. Numbers reflect hours after transfection. (Scale bar, 15 μm). (Adapted with permission from Tsvetkov et al. 2010)

Thus, the relationship between the capacity of the system and the load that is placed on it becomes the arbiter of whether that neuron is fated to degenerate or not (Fig. 6). In this model, genetic mutations (e.g., those that cause HD or familial forms of ALS, PD, or AD) or the as yet ill-defined environmental influences (e.g., those that presumably contribute to sporadic ALS, PD, and AD) work by shifting the balance toward neurodegenerative disease by increasing the production of proteins prone to misfold, by hampering their clearance, or both. In at least some

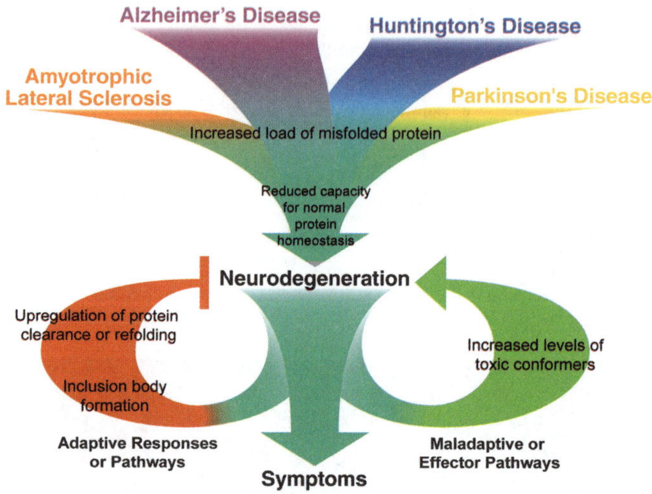

Fig. 6 Protein homeostasis may play a critical role in many adult onset neurodegenerative diseases. In this model, these diseases have in common an increase in the production and/or a decrease in the clearance of misfolded protein, leading to an overall increase in the load of misfolded protein. In turn, this additional load of misfolded protein reduces the capacity of cells to re-fold or clear other misfolded proteins that are generated during the course of normal physiology. The systemic stress may induce adaptive changes, such as IB formation or an upregulation of pathways that re-fold or clear misfolded proteins and mitigate neurodegeneration. However, the imbalance between the load of misfolded proteins and the cellular capacity to deal with them may place further stress on the re-folding or clearance machinery and lead to even greater levels of toxic protein conformers. Ultimately, a tonic imbalance could lead to chronic widespread misfolding of other important proteins and to neuronal dysfunction or death and, eventually, symptoms

cases, such as HD, individuals express the disease-causing mutation from birth and yet four decades may pass before they become overtly symptomatic. This delay suggests a role for aging in the production or clearance of misfolded proteins. Alternatively, these diseases may be characterized by a life-long modest mismatch in protein homeostasis and a state of tonic low-level neuronal dysfunction. Over decades, the body's coping mechanisms may become exhausted, triggering a cascade of systems failures that culminate in symptoms.

Is this the only common thread in neurodegenerative disease? Probably not, but it is an excellent example of how we need to look beyond simple pathology to guide the search for new treatments for these devastating diseases.

Acknowledgments This work was supported with grants from the National Institute of Neurological Disease and Stroke (NS 39074 and NS45491), National Institute of Aging (AG022074), Taube-Koret Center Huntington's Disease Research, the Hellman Family Foundation Program in Alzheimer's Disease Research, the Hereditary Disease Foundation, and the Huntington's Disease Society of America.

References

Alves-Rodrigues A, Gregori L, Figueiredo-Pereira ME (1998) Ubiquitin, cellular inclusions and their role in neurodegeneration. Trends Neurosci 21:516–520

Arrasate M, Finkbeiner S (2005) Automated microscope system for determining factors that predict neuronal fate. Proc Natl Acad Sci USA 102:3840–3845

Arrasate M, Mitra S, Schweitzer ES, Segal MR, Finkbeiner S (2004) Inclusion body formation reduces levels of mutant huntingtin and the risk of neuronal death. Nature 431:805–810

Balch WE, Morimoto RI, Dillin A, Kelly JW (2008) Adapting proteostasis for disease intervention. Science 319:916–919

Barmada SJ, Skibinski G, Korb E, Rao EJ, Wu JY, Finkbeiner S (2010) Cytoplasmic mislocalization of TDP-43 is toxic to neurons and enhanced by a mutation associated with familial amyotrophic lateral sclerosis. J Neurosci 30:639–649

Beal MF (2005) Mitochondria take center stage in aging and neurodegeneration. Ann Neurol 58:495–505

Beal FM, Bossy-Wetzel E, Finkbeiner S, Fiskum G, Glasson B, Johnson C, Khachaturian ZS, Lee VMY, Nicholls D, Reddy H, Reynolds I, Teplow DB, Thal LJ, Trojanowski JQ, Walsh DM, Wetzel R, Wexler NS, Young AB, Bain L (2006) Common threads in neurodegenerative disorders of aging. Alzheimers Dement 2:322–326

Becher MW, Kotzuk JA, Sharp AH, Davies SW, Bates GP, Price DL, Ross CA (1998) Intranuclear neuronal inclusions in Huntington's disease and Dentatorubral and Pallidoluysian atrophy – Correlation between the density of inclusions and IT-15 CAG triplet repeat length. Neurobiol Dis 4:387–397

Bedford L, Hay D, Devoy A, Paine S, Powe DG, Seth R, Gray T, Topham I, Fone K, Rezvani N, Mee M, Soane T, Layfield R, Sheppard PW, Ebendal T, Usoskin D, Lowe J, Mayer RJ (2008) Depletion of 26S proteasomes in mouse brain neurons causes neurodegeneration and Lewy-like inclusions resembling human pale bodies. J Neurosci 28:8189–8198

Brooks E, Arrasate M, Cheung K, Finkbeiner SM (2004) Using antibodies to analyze polyglutamine stretches. Methods Mol Biol 277:103–128

Dagda R, Zhu J, Kulich S, Chu C (2008) Mitochondrially localized ERK2 regulates mitophagy and autophagic cell stress: implications for Parkinson's disease. Autophagy 4:770–782

Daub A, Sharma P, Finkbeiner S (2009) High-content screening of primary neurons: ready for prime time. Curr Opin Neurobiol 19:537–543

Davies SW, Turmaine M, Cozens BA, DiFiglia M, Sharp AH, Ross CA, Scherzinger E, Wanker EE, Mangiarini L, Bates GP (1997) Formation of neuronal intranuclear inclusions underlies the neurological dysfunction in mice transgenic for the HD mutation. Cell 90:537–548

Fink AL (1998) Protein aggregation: folding aggregates, inclusion bodies and amyloid. Fold Des 3:R9–R23

Finkbeiner S, Cuervo AM, Morimoto RI, Muchowski PJ (2006) Disease-modifying pathways in neurodegeneration. J Neurosci 26:10349–10357

Floyd JA, Hamilton BA (1999) Intranuclear inclusions and the ubiquitin-proteasome pathway: digestion of a red herring? Neuron 24:765–766

Goedert M, Spillantini MG, Davies SW (1998) Filamentous nerve cell inclusions in neurodegenerative diseases. Curr Opin Neurobiol 8:619–632

Humbert S, Bryson EA, Cordelières FP, Connors NC, Datta SR, Finkbeiner S, Greenberg ME, Saudou F (2002) The IGF-1/Akt pathway is neuroprotective in Huntington's disease and involves huntingtin phosphorylation by Akt. Dev Cell 3:1–20

Jagadha V, Becker LE (1989) Dendritic pathology: an overview of Golgi studies in man. Can J Neurol Sci 16:41–50

Kanki T, Wang K, Cao Y, Baba M, Klionsky DJ (2009) Atg32 is a mitochondrial protein that confers selectivity during mitophagy. Dev Cell 17:98–109

Klement IA, Skinner PJ, Kaytor MD, Yi H, Hersch SM, Clark HB, Zoghbi HY, Orr HT (1998) Ataxin-1 nuclear localization and aggregation: role in polyglutamine-induced disease in SCA1 transgenic mice. Cell 95:41–53

Maccioni RB, Farias G, Morales I, Navarrete L (2010) The revitalized tau hypothesis on Alzheimer's disease. Arch Med Res 41:226–231

Mackenzie IRA, Baker M, Pickering-Brown S, Hsiung G-YR, Lindholm C, Dwosh E, Gass J, Cannon A, Rademakers R, Hutton M, Feldman HH (2006) The neuropathology of frontotemporal lobar degeneration caused by mutations in the *progranulin* gene. Brain 129:3081–3090

Miller J, Arrasate M, Shaby BA, Mitra S, Masliah E, Finkbeiner S (2010) Quantitative relationships between huntingtin levels, polyglutamine length, inclusion body formation, and neuronal death provide novel insight into Huntington's disease molecular pathogenesis. J Neurosci 30:10541–10550

Mitra S, Finkbeiner S (2008) The ubiquitin-proteasome pathway in Huntington's disease. ScientificWorldJournal 8:421–433

Mitra S, Tsvetkov AS, Finkbeiner S (2009) Single neuron ubiquitin-proteasome dynamics accompanying inclusion body formation in Huntington disease. J Biol Chem 284:4398–4403

Mura E, Lanni C, Preda S, Pistoia F, Sara M, Racchi M, Schettini G, Marchi M, Govoni S (2010) β-amyloid: A disease target or a synaptic regulator affecting age-related neurotransmitter changes? Curr Pharm Des 16:672–683

Nakamura K, Nemani VM, Azarbal F, Skibinski G, Levy JM, Egami K, Munishkina L, Zhang J, Gardner B, Wakabayashi J, Sesaki H, Cheng Y, Finkbeiner S, Nussbaum RL, Masliah E, Edwards RH (2011) Direct membrane association drives mitochondrial fission by the Parkinson Disease-associated protein α-synuclein. J Biol Chem 286:20710–20726

Pamphlett R, Luquin N, McLean C, Jew SK, Adams L (2009) TDP-43 neuropathology is similar in sporadic amyotrophic lateral sclerosis with or without TDP-43 mutations. Neuropathol Appl Neurobiol 35:222–225

Peters-Libeu C, Newhouse Y, Krishnan P, Cheung K, Brooks E, Weisgraber K, Finkbeiner S (2005) Crystallization and diffraction properties of the Fab fragment of 3B5H10, an antibody specific for disease-causing polyglutamine stretches. Acta Crystallogr Sect F Struct Biol Cryst Commun 61:1065–1068

Powers ET, Morimoto RI, Dillin A, Kelly JW, Balch WE (2009) Biological and chemical approaches to diseases of proteostasis deficiency. Annu Rev Biochem 78:959–991

Ramdzan YM, Nisbet RM, Miller J, Finkbeiner S, Hill AF, Hatters DM (2010) Conformation sensors that distinguish monomeric proteins from oligomers in live cells. Chem Biol 17:371–379

Robakis NK (2010) Are Aβ and its derivatives causative agents or innocent bystanders in AD? Neurodegener Dis 7:32–37

Robitaille Y, Lopes-Cendes I, Becher M, Rouleau G, Clark AW (1997) The neuropathology of CAG repeat diseases: review and update of genetic and molecular features. Brain Pathol 7:901–926

Saudou F, Finkbeiner S, Devys D, Greenberg ME (1998) Huntingtin acts in the nucleus to induce apoptosis, but death does not correlate with the formation of intranuclear inclusions. Cell 95:55–66

Scheff SW, Price DA (2003) Synaptic pathology in Alzheimer's disease: a review of ultrastructural studies. Neurobiol Aging 24:1029–1046

Shacka JJ, Roth KA, Zhang J (2008) The autophagy-lysosomal degradation pathway: role in neurodegenerative disease and therapy. Front Biosci 13:718–736

Sisodia SS (1998) Nuclear inclusions in glutamine repeat disorders: are they pernicious, coincidental or beneficial? Cell 95:1–4

Struble RG, Ala T, Patrylo PR, Brewer GJ, Yan XX (2010) Is brain amyloid production a cause or a result of dementia of the Alzheimer's type? J Alzheimers Dis 22:393–399

Tsvetkov AS, Mitra S, Finkbeiner S (2009) Protein turnover differences between neurons and other cells. Autophagy 5:1–3

Tsvetkov AS, Miller J, Arrasate M, Wong JS, Pleiss MA, Finkbeiner S (2010) A small-molecule scaffold induces autophagy in primary neurons and protects against toxicity in a Huntington disease model. Proc Natl Acad Sci USA 107:16982–16987

Zhang QC, Yeh TL, Leyva A, Frank LG, Miller J, Kim YE, Langen R, Finkbeiner S, Amzel ML, Ross CA, Poirier MA (2011) A compact beta model of huntingtin toxicity. J Biol Chem 286:8188–8196

Therapeutic potential of longevity modulators as neuroprotective targets in neurodegenerative disease

Rafael Vazquez-Manrique, Cendrine Tourette, and Christian Neri

Abstract Several genes and pathways may act in a concerted and finely tuned fashion to regulate cell survival and longevity. Pathways that converge onto FOXO transcription factors have been the subject of particular attention because FOXO is a longevity-promoting factor that may be important throughout the entire lifetime of a living organism, as suggested by its role in stem cell homeostasis, cell differentiation and adult cell survival. Additionally, there is accumulating evidence on protective activities by FOXO and its interactors in age-associated diseases such as metabolic disorder and neurodegeneration. This review briefly summarizes the findings suggesting that key genetic modulators of life span such as FOXO and its interactors constitute promising targets to develop neurodegenerative disease-modifying strategies.

Introduction

A major challenge in neurodegenerative disease research is to identify disease-modifying strategies. Given the biological and clinical complexity of sporadic neurodegenerative disease such as Alzheimer's disease (AD) or Parkinson's disease (PD), the study of inherited neurodegenerative disease is expected to shed light on the mechanisms that underlie neurodegeneration. In this respect, Huntington's disease (HD), a dominantly inherited disease with CAG expansion in the huntingtin (htt) gene and expanded polyglutamine (polyQ) tracts in the htt protein causing striatal and cortical degeneration (Walker 2007), has become a 'model neurodegenerative disease.' The large number of models for HD, the knowledge

---

R. Vazquez-Manrique • C. Neri (✉)
Neuronal Cell Biology and Pathology, INSERM Unit 894, 75014 Paris, France
e-mail: christian.neri@inserm.fr

C. Tourette
Buck Institute, Novato, CA 94945, USA

R.I. Morimoto and Y. Christen (eds.), *Protein Quality Control in Neurodegenerative Diseases*, Research and Perspectives in Alzheimer's Disease,
DOI 10.1007/978-3-642-27928-7_9,

accumulated on HD mechanisms (Zuccato et al. 2010) and the availability of well-characterized cohorts of HD subjects indeed provide a unique framework in which to investigate the therapeutic potential of specific target genes and pathways. Among the many genes and mechanisms that could be targeted for therapeutic intervention in HD (Zuccato et al. 2010), longevity modulators and, more largely, genes that are essential to cellular homeostasis are the subject of intense research because they may directly regulate a wide range of important cell survival mechanisms such as, for example, oxidative stress response and protein quality control. While several genes may modulate life span (Narasimhan et al. 2009), a limited number of genes may function to increase longevity. This is notably true for the FOXO proteins, a family of forkhead transcription factors that is conserved from *C. elegans* to mammals. FOXO transcription factors indeed increase longevity in invertebrates, and in mammals they have a wide range of functions, as further developed below.

The FOXO network

Several signaling pathways — such as the PI3K-AKT pathway, stress-activated pathways such as the MST-1 and JNK pathways, and the SIR-2 and AMPK pathways — converge onto FOXO to regulate a variety of biological processes that are important to development, metabolism, tumor suppression and longevity (Calnan and Brunet 2008). A large number of studies have indicated that, in response to external stimuli including growth factors such as insulin and insulin-like growth factor (IGF-1), nutrients, cytokines and oxidative stress stimuli, these pathways are able to regulate the subcellular localization, DNA binding activity, transcriptional activity and stability of these FOXO proteins. These studies have been thoroughly reviewed in excellent articles (Calnan and Brunet 2008; Landis and Murphy 2010; Yen et al. 2011); a finely tuned regulatory network is emphasized in which, through a series of context-dependent post-translational modifications and interactions with co-regulators, FOXO is able to integrate several types of external stimuli to regulate gene expression and modulate cell cycle arrest, apoptosis, autophagy, angiogenesis, differentiation, stress resistance, stem cell maintenance, glucogenesis, and food intake. Nutrient-sensing pathways appear to be important upstream components of this regulatory network. Nutrient-sensing factors such as insulin/insulin-like growth factor (IGF) signaling (IIS), NAD^+-dependent histone deacetylases (sirtuins/SIR-2/SIRTs), the target of rapamycin (TOR) kinase and ribosomal S6 Kinase (TOR/S6K) pathway and the AMPK pathway may be targeted by dietary restriction to slow aging (Cabreiro and Gems 2010; Chiacchiera and Simone 2010). Additionally, nutrient-sensing pathways are key regulators of FOXO activity. For example, it is well-established that IIS modifies FOXO at specific phosphorylation sites, with reduction of IIS activating FOXO, which may translate into life span extension in invertebrates (Calnan and Brunet 2008). Besides IIS and modulation of FOXO phosphorylation, the SIRT1 nutrient-sensing pathway and modulation of FOXO acetylation is another important FOXO regulator. Increased *sir-2.1*/SIRT1 dosage in high-copy *sir-2.1* transgene-containing strains was indeed found to significantly increase life span in *C. elegans*, an effect that requires *daf-16*/FOXO (Tissenbaum and Guarente 2001), suggesting that sirtuins could play important roles in life span extension. This notion has been recently challenged by additional studies in *C. elegans* and *Drosophila melanogaster* (Burnett et al. 2011; Rizki et al. 2011), which revealed that the strong level of life span extension may be the result of unrelated mutations (here mutations producing a dye-filling, defective/Dyf phenotype) in the high-copy *sir-2.1* overexpressor strains initially used to study longevity-promoting effects (Burnett et al. 2011; Rizki et al. 2011). These studies raised the possibility that sirtuins might not be longevity-promoting factors to that extent. However, at the same time, the use of low-copy *sir-2.1* transgene-containing strains consistently indicated that a significant level (10-15 %) of life span extension may be achieved by increased *sir-2.1* dosage, as reported by two independent studies (Rizki et al. 2011; Viswanathan and Guarente 2011). Importantly, the results of one of these studies (Rizki et al. 2011) were consistent with the notion that *sir-2.1* interacts with *daf-16*/FOXO and its cofactors such as 14-3-3 proteins in modulating life span (Berdichevsky et al. 2006). As further discussed below, unrelated mutations were not responsible for the strong level of neuroprotection achieved against mutant polyQs by *sir-2.1* overexpression (Burnett et al. 2011). Thus, while additional investigation will be needed to determine to what extent sirtuins may be able to promote life span extension, sirtuins may remain interesting factors for their role in promoting longevity. They also appear to stand as one key example of the relevance of longevity modulators to age-related diseases such as metabolic disease and neurodegenerative disease (Kim et al. 2007; Baur et al. 2010).

In addition to being regulated by kinases and deacetylases, FOXO activity is regulated by co-factors such as ß-catenin. The ß-catenin protein was indeed found to bind to FOXO and to regulate its transcriptional activity in response to oxidative stress (Essers et al. 2005). Additionally, FOXO may compete with the canonical Wnt effector T cell factor (TCF) for binding to ß-catenin (Hoogeboom et al. 2008), which further highlights the importance of ß-catenin in the regulation of FOXO activity. Thus, while FOXO may act together with other transcription factors to regulate gene expression (Landis and Murphy 2010), FOXO appears to stand at the center of a large network of upstream regulators, co-factors and downstream targets (Fig. 1) that are important for cell homeostasis and survival. Interestingly, studies based on chromatin immunoprecipitation (Oh et al. 2006; Schuster et al. 2010), gene expression profiling in *C. elegans* (Murphy et al. 2003) and mouse neural stem cells (Paik et al. 2009; Renault et al. 2009) and other approaches (Lee et al. 2003; Yu et al. 2008) suggested that thousands of genes might be regulated by FOXO factors (Jensen et al. 2006; Murphy 2006). These results open an avenue to a better understanding of FOXO activity in regulating downstream mechanisms in specific contexts. Given that FOXO may act in specific cells and tissues to execute different outputs (Landis and Murphy 2010), it is anticipated that specific transcriptional targets may be engaged in the different responses that are mediated by FOXO.

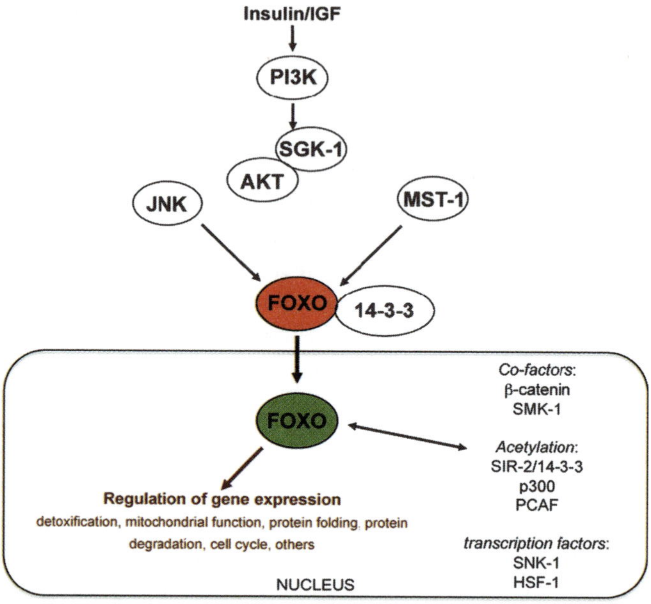

Fig. 1 Simplified view of the FOXO network and its major players

Interestingly, in addition to being important for stem cell homeostasis and adult longevity, FOXO factors are involved in development and cell differentiation (Kitamura et al. 2007; Demontis and Perrimon 2009; Yuan et al. 2009; de la Torre-Ubieta et al. 2010; Kerdiles et al. 2010; Schuff et al. 2010), and they may be important in promoting adult neurogenesis, as suggested by studies in Drosophila (Siegrist et al. 2010). Thus, FOXO factors may be essential to cell and tissue homeostasis during the entire lifetime of a living organism. As such, it is not surprising that FOXO and its main regulators such as sirtuins have been implicated in the protection against age-associated diseases such as cancer, diabetes, cardiovascular disease and neurodegeneration (Dansen and Burgering 2008; Chiacchiera and Simone 2010; Herranz and Serrano 2010).

The SIR-2/FOXO pathway and its role in degenerative disease

For the same reason that pathways converging onto FOXO are important to cell survival and longevity, they may be of particular interest in exploring how to promote neuron or muscle cell survival in human degenerative disease. Over the past years, accumulating evidence has been obtained on the significant potential of FOXO and its interactors for age-related diseases. Perhaps the first observation suggesting that longevity-promoting factors might be able to modulate the toxicity of degenerative disease proteins was obtained in the nematode *C. elegans*. In transgenic nematodes expressing polyQ tracts in body wall muscle cells, reducing IIS by use of *age-1*/PI3K loss-of-function (LOF) mutants decreased the toxicity associated with expanded polyQs, and this effect was dependent on *daf-16*/FOXO (Morley et al. 2002). Another interesting observation was obtained in a study based on the Wallerian degeneration slow mice, which showed that NAD and SIRT1 promoted axonal protection in these mice (Araki et al. 2004). Direct evidence for longevity-promoting factors such as SIR-2 and FOXO to protect diseased neurons at an early stage of cell decline was obtained by using a *C. elegans* model of HD pathogenesis in which expression of polyQ-expanded N-terminal huntingtin in touch receptor mechanosensory neurons induced neuronal dysfunction (loss of response to light touch; Parker et al. 2001). This behavioral phenotype reflected the early stages of mutant polyQ toxicity, as it was accompanied by axonal dystrophy and occurred in the absence of cell death (Parker et al. 2001). In these animals, increased *sir-2.1*/SIRT1 dosage protects from expanded polyQs, with reduction of axonal dystrophy, whereas *sir-2.1* LOF aggravates expanded-polyQ toxicity (Parker et al. 2005). Additionally, *sir-2.1* neuroprotection in this model requires *daf-16*/FOXO (Parker et al. 2005), which emphasizes a role for the SIR-2.1/DAF-16 pathway in early-stage neuronal protection from expanded polyQs and suggests that raising the activity of the SIRT1/FOXO pathway may protect from mutant huntingtin in mammalian models of HD. Importantly, this neuroprotective effect is not due to unrelated mutations (i.e., mutations producing a Dyf phenotype) in the high-copy *sir-2.1* transgene containing strains, as neuroprotection was observed to the same extent in non-Dyf, *sir-2.1* overexpressors (Burnett et al. 2011). Consistent with the notion that SIR-2 may protect neurons from the early stage cytotoxicity induced by neurodegenerative disease-associated proteins, neuroprotection by *sir-2.1* overexpression was also demonstrated in transgenic *C. elegans* nematodes expressing octa-repeat expanded PrP in mechanosensory neurons (Bizat et al. 2010). In these animals, the neuroprotective effect of *sir-2.1* was unrelated to an effect on the conformation of mutant PrP, since *sir-2.1* overexpression did not reduce proteinase K-resistant PrP levels (Bizat et al. 2010). Several studies have further extended the neuroprotective activity of SIR-2/SIRT1 and FOXO in models of neurodegenerative diseases. Sirtuin SIRT1 was upregulated in mouse models of AD and amyotrophic lateral sclerosis (ALS), and SIRT1 was protective in cell-based models of ALS and AD/taupathies (Kim et al. 2007), providing evidence of SIRT1 protecting diseased neurons. Neurite outgrowth, cell viability and tolerance to degeneration induced by Aß(1–42) oligomers were promoted in primary neurons from mice with neuron-specific SIRT1 overexpression (Guo et al. 2011), further supporting the notion that SIRT1 is able to protect diseased neurons. In transgenic *C. elegans* nematodes expressing Aß(1–42) in body wall muscle cells (Link et al. 2003), *daf-16*/FOXO and heat shock factor 1 (*hsf-1*) cooperated to regulate a two-step cell survival response involving disaggregation/aggregation (Cohen et al. 2006). Consistently, in a mouse model of AD (APPswe/Presenilin1ΔE9 mice), genetically reducing IGF signaling protected from Alzheimer's-like disease symptoms (Cohen et al. 2009). Finally, allelic variation in FOXO3A appears to be consistently associated with the ability to be long-lived, as suggested by several association studies in populations of centenarians (Willcox et al. 2008; Anselmi et al. 2009; Flachsbart et al. 2009; Li et al. 2009; Soerensen et al. 2010; Kleindorp et al. 2011).

While the SIR-2.1/DAF-16 pathway may protect neurons from mutant polyQ cytotoxicity, it actually showed opposite effects in a *C. elegans* model of oculopharyngeal muscular dystrophy (OPMD), a late-onset disease caused by expanded polyalanines in the PABPN1 protein (Catoire et al. 2008). In *C. elegans* transgenics expressing human PABPN1 in body wall muscle cells, *sir-2.1* overexpression indeed aggravated muscle cell decline and abnormal motility caused by mutant PABPN1 expression, whereas *sir-2.1* LOF was protective in a *daf-16*/FOXO-dependent manner (Catoire et al. 2008). These observations suggest that the status of FOXO activity in mutant polyQ neurons is different from that of mutant polyAla muscle cells. These results also suggested that SIRT1 inhibitors have therapeutic potential in OPMD, as further illustrated by pharmacological studies in PABPN1 nematodes (Pasco et al. 2010b). Furthermore, consistent with the notion that ß-catenin signaling is a pro-survival pathway that may act as an important regulator of FOXO activity (Essers et al. 2005), a genetic link between the SIR-2/FOXO pathway and canonical Wnt pathway was identified for protection against mutant PABPN1 toxicity (Pasco et al. 2010a).

Over the past years, studies of SIRT1/FOXO activity in models of disease have fostered the search for SIRT1 activators and inhibitors with clinical potential. While we do not aim here to review drug discovery and development in the field of sirtuin research, it is important to note that the mechanisms of action of SIRT1-activating compounds (STACs) have been the subject of debate. This applies to resveratrol, a plant polyphenol that has a broad range of beneficial activities on health span (Wood et al. 2004; Baur et al. 2006; Fulda and Debatin 2006; Lagouge et al. 2006), is neuroprotective in several models of neurodegenerative diseases (Parker et al. 2005; Kim et al. 2007; Bizat et al. 2010; Richard et al. 2011) and has attracted a lot of attention as a potential drug (Patel et al. 2011). Resveratrol was initially thought to primarily activate SIRT1 (Howitz et al. 2003). It is now thought that resveratrol has multiple mechanisms of action (Harikumar and Aggarwal 2008) as, besides indirect SIRT1 activation, it may use other targets such as, for example, AMPK (Dasgupta and Milbrandt 2007) and peroxisome proliferator-activated receptor coactivator-1alpha (PGC-1alpha; Lagouge et al. 2006). This finding promoted the search for selective SIRT1 activators (Milne et al. 2007), which resulted in the characterization of their biological properties as caloric restriction mimetics (Smith et al. 2009). However, while STACs hold promise for the treatment of age-associated diseases, their mechanisms of action and therapeutic potential are controversial, as with the situation encountered with resveratrol. Some studies raised the possibility that the complex formation between STACs and fluorophore-labeled substrate peptides played a role in the activation of SIRT1 (Pacholec et al. 2010), whereas other studies subsequently showed that these compounds were able to activate unlabelled peptides, suggesting that STACs interact directly with SIRT1 and activate SIRT1-catalyzed deacetylation through an allosteric mechanism (Dai et al. 2010). Further developments in this exciting field of research will be needed to provide more insights into the mechanisms of action and therapeutical potential of STACs.

Perspectives

As highlighted above, there is accumulating evidence about the protective role of FOXO and its interactors in age-related diseases. Interestingly, FOXO may also be involved in the regulation of normal age-related processes, as suggested by its effect on age-related neurite branching in *C. elegans* (Tank et al. 2011). One of the current challenges is to identify the mechanisms that, downstream to FOXO, are key to neuron survival in neurodegenerative disease. Given the complex regulation of gene expression by FOXO and its interactors, it will be important to investigate how FOXO transcriptional targets may be involved in the regulation of survival responses in response to a specific neurodegenerative disease protein and to evaluate whether this may be relevant to other neurodegenerative diseases. At the methodological level, integrating genetics, cell biology and quantitative biology is likely to provide new and important insights into the biological significance and therapeutic potential of the FOXO network.

References

Anselmi CV, Malovini A, Roncarati R, Novelli V, Villa F, Condorelli G, Bellazzi R, Puca AA (2009) Association of the FOXO3A locus with extreme longevity in a southern Italian centenarian study. Rejuvenation Res 12:95–104

Araki T, Sasaki Y, Milbrandt J (2004) Increased nuclear NAD biosynthesis and SIRT1 activation prevent axonal degeneration. Science 305:1010–1013

Baur JA, Pearson KJ, Price NL, Jamieson HA, Lerin C, Kalra A, Prabhu VV, Allard JS, Lopez-Lluch G, Lewis K, Pistell PJ, Poosala S, Becker KG, Boss O, Gwinn D, Wang M, Ramaswamy S, Fishbein KW, Spencer RG, Lakatta EG, Le Couteur D, Shaw RJ, Navas P, Puigserver P, Ingram DK, de Cabo R, Sinclair DA (2006) Resveratrol improves health and survival of mice on a high-calorie diet. Nature 444:337–342

Baur JA, Chen D, Chini EN, Chua K, Cohen HY, de Cabo R, Deng C, Dimmeler S, Gius D, Guarente LP, Helfand SL, Imai S, Itoh H, Kadowaki T, Koya D, Leeuwenburgh C, McBurney M, Nabeshima Y, Neri C, Oberdoerffer P, Pestell RG, Rogina B, Sadoshima J, Sartorelli V, Serrano M, Sinclair DA, Steegborn C, Tatar M, Tissenbaum HA, Tong Q, Tsubota K, Vaquero A, Verdin E (2010) Dietary restriction: standing up for sirtuins. Science 329:1012–1013

Berdichevsky A, Viswanathan M, Horvitz HR, Guarente L (2006) *C. elegans* SIR-2.1 interacts with 14-3-3 proteins to activate DAF-16 and extend life span. Cell 125:1165–1177

Bizat N, Peyrin JM, Haik S, Cochois V, Beaudry P, Laplanche JL, Neri C (2010) Neuron dysfunction is induced by prion protein with an insertional mutation via a Fyn kinase and reversed by sirtuin activation in Caenorhabditis elegans. J Neurosci 30:5394–5403

Burnett C, Valentini S, Cabreiro F, Goss M, Somogyvári M, Piper MD, Hoddinott M, Sutphin GL, Leko V, McElwee JJ, Vazquez-Manrique RP, Orfila AM, Ackerman D, Au C, Vinti G, Riesen M, Howard K, Neri C, Bedalov A, Kaeberlein M, Soti C, Partridge L, Gems D (2011) Absence of effects of Sir2 overexpression on lifespan in *C. elegans* and Drosophila. Nature 477:482–485

Cabreiro F, Gems D (2010) Treating aging: progress toward dietary restriction mimetics. F1000 Biol Rep 2:76

Calnan DR, Brunet A (2008) The FoxO code. Oncogene 27:2276–2288

Catoire H, Pasco MY, Abu-Baker A, Holbert S, Tourette C, Brais B, Rouleau GA, Parker JA, Neri C (2008) Sirtuin inhibition protects from the polyalanine muscular dystrophy protein PABPN1. Hum Mol Genet 17:2108–2117

Chiacchiera F, Simone C (2010) The AMPK-FoxO3A axis as a target for cancer treatment. Cell Cycle 9:1091–1096

Cohen E, Bieschke J, Perciavalle RM, Kelly JW, Dillin A (2006) Opposing activities protect against age-onset proteotoxicity. Science 313:1604–1610

Cohen E, Paulsson JF, Blinder P, Burstyn-Cohen T, Du D, Estepa G, Adame A, Pham HM, Holzenberger M, Kelly JW, Masliah E, Dillin A (2009) Reduced IGF-1 signaling delays age-associated proteotoxicity in mice. Cell 139:1157–1169

Dai H, Kustigian L, Carney D, Case A, Considine T, Hubbard BP, Perni RB, Riera TV, Szczepankiewicz B, Vlasuk GP, Stein RL (2010) SIRT1 activation by small molecules: kinetic and biophysical evidence for direct interaction of enzyme and activator. J Biol Chem 285:32695–32703

Dansen TB, Burgering BM (2008) Unravelling the tumor-suppressive functions of FOXO proteins. Trends Cell Biol 18:421–429

Dasgupta B, Milbrandt J (2007) Resveratrol stimulates AMP kinase activity in neurons. Proc Natl Acad Sci USA 104:7217–7222

de la Torre-Ubieta L, Gaudilliere B, Yang Y, Ikeuchi Y, Yamada T, DiBacco S, Stegmuller J, Schuller U, Salih DA, Rowitch D, Brunet A, Bonni A (2010) A FOXO-Pak1 transcriptional pathway controls neuronal polarity. Genes Dev 24:799–813

Demontis F, Perrimon N (2009) Integration of Insulin receptor/Foxo signaling and dMyc activity during muscle growth regulates body size in Drosophila. Development 136:983–993

Essers MA, de Vries-Smits LM, Barker N, Polderman PE, Burgering BM, Korswagen HC (2005) Functional interaction between beta-catenin and FOXO in oxidative stress signaling. Science 308:1181–1184

Flachsbart F, Caliebe A, Kleindorp R, Blanche H, von Eller-Eberstein H, Nikolaus S, Schreiber S, Nebel A (2009) Association of FOXO3A variation with human longevity confirmed in German centenarians. Proc Natl Acad Sci USA 106:2700–2705

Fulda S, Debatin KM (2006) Resveratrol modulation of signal transduction in apoptosis and cell survival: a mini-review. Cancer Detect Prev 30:217–223

Guo W, Qian L, Zhang J, Zhang W, Morrison A, Hayes P, Wilson S, Chen T, Zhao J (2011) Sirt1 overexpression in neurons promotes neurite outgrowth and cell survival through inhibition of the mTOR signaling. J Neurosci Res 89:1723–1736

Harikumar KB, Aggarwal BB (2008) Resveratrol: a multitargeted agent for age-associated chronic diseases. Cell Cycle 7:1020–1035

Herranz D, Serrano M (2010) SIRT1: recent lessons from mouse models. Nat Rev Cancer 10:819–823

Hoogeboom D, Essers MA, Polderman PE, Voets E, Smits LM, Burgering BM (2008) Interaction of FOXO with beta-catenin inhibits beta-catenin/T cell factor activity. J Biol Chem 283:9224–9230

Howitz KT, Bitterman KJ, Cohen HY, Lamming DW, Lavu S, Wood JG, Zipkin RE, Chung P, Kisielewski A, Zhang LL, Scherer B, Sinclair DA (2003) Small molecule activators of sirtuins extend Saccharomyces cerevisiae lifespan. Nature 425:191–196

Jensen VL, Gallo M, Riddle DL (2006) Targets of DAF-16 involved in Caenorhabditis elegans adult longevity and dauer formation. Exp Gerontol 41:922–927

Kerdiles YM, Stone EL, Beisner DR, McGargill MA, Ch'en IL, Stockmann C, Katayama CD, Hedrick SM (2010) Foxo transcription factors control regulatory T cell development and function. Immunity 33:890–904

Kim D, Nguyen MD, Dobbin MM, Fischer A, Sananbenesi F, Rodgers JT, Delalle I, Baur JA, Sui G, Armour SM, Puigserver P, Sinclair DA, Tsai LH (2007) SIRT1 deacetylase protects against neurodegeneration in models for Alzheimer's disease and amyotrophic lateral sclerosis. EMBO J 26:3169–3179

Kitamura T, Kitamura YI, Funahashi Y, Shawber CJ, Castrillon DH, Kollipara R, DePinho RA, Kitajewski J, Accili D (2007) A Foxo/Notch pathway controls myogenic differentiation and fiber type specification. J Clin Invest 117:2477–2485

Kleindorp R, Flachsbart F, Puca AA, Malovini A, Schreiber S, Nebel A (2011) Candidate gene study of FOXO1, FOXO4, and FOXO6 reveals no association with human longevity in Germans. Aging Cell 10:622–628

Lagouge M, Argmann C, Gerhart-Hines Z, Meziane H, Lerin C, Daussin F, Messadeq N, Milne J, Lambert P, Elliott P, Geny B, Laakso M, Puigserver P, Auwerx J (2006) Resveratrol improves mitochondrial function and protects against metabolic disease by activating SIRT1 and PGC-1alpha. Cell 127:1109–1122

Landis JN, Murphy CT (2010) Integration of diverse inputs in the regulation of Caenorhabditis elegans DAF-16/FOXO. Dev Dyn 239:1405–1412

Lee SS, Kennedy S, Tolonen AC, Ruvkun G (2003) DAF-16 target genes that control *C. elegans* life-span and metabolism. Science 300:644–647

Li Y, Wang WJ, Cao H, Lu J, Wu C, Hu FY, Guo J, Zhao L, Yang F, Zhang YX, Li W, Zheng GY, Cui H, Chen X, Zhu Z, He H, Dong B, Mo X, Zeng Y, Tian XL (2009) Genetic association of FOXO1A and FOXO3A with longevity trait in Han Chinese populations. Hum Mol Genet 18:4897–4904

Link CD, Taft A, Kapulkin V, Duke K, Kim S, Fei Q, Wood DE, Sahagan BG (2003) Gene expression analysis in a transgenic Caenorhabditis elegans Alzheimer's disease model. Neurobiol Aging 24:397–413

Milne JC, Lambert PD, Schenk S, Carney DP, Smith JJ, Gagne DJ, Jin L, Boss O, Perni RB, Vu CB, Bemis JE, Xie R, Disch JS, Ng PY, Nunes JJ, Lynch AV, Yang H, Galonek H, Israelian K, Choy W, Iffland A, Lavu S, Medvedik O, Sinclair DA, Olefsky JM, Jirousek MR, Elliott PJ, Westphal CH (2007) Small molecule activators of SIRT1 as therapeutics for the treatment of type 2 diabetes. Nature 450:712–716

Morley JF, Brignull HR, Weyers JJ, Morimoto RI (2002) The threshold for polyglutamine-expansion protein aggregation and cellular toxicity is dynamic and influenced by aging in Caenorhabditiselegans. Proc Natl Acad Sci USA 16:16

Murphy CT (2006) The search for DAF-16/FOXO transcriptional targets: approaches and discoveries. Exp Gerontol 41:910–921

Murphy CT, McCarroll SA, Bargmann CI, Fraser A, Kamath RS, Ahringer J, Li H, Kenyon C (2003) Genes that act downstream of DAF-16 to influence the lifespan of Caenorhabditis elegans. Nature 424:277–283

Narasimhan SD, Yen K, Tissenbaum HA (2009) Converging pathways in lifespan regulation. Curr Biol 19:R657–R666

Oh SW, Mukhopadhyay A, Dixit BL, Raha T, Green MR, Tissenbaum HA (2006) Identification of direct DAF-16 targets controlling longevity, metabolism and diapause by chromatin immunoprecipitation. Nat Genet 38:251–257

Pacholec M, Bleasdale JE, Chrunyk B, Cunningham D, Flynn D, Garofalo RS, Griffith D, Griffor M, Loulakis P, Pabst B, Qiu X, Stockman B, Thanabal V, Varghese A, Ward J, Withka J, Ahn K (2010) SRT1720, SRT2183, SRT1460, and resveratrol are not direct activators of SIRT1. J Biol Chem 285:8340–8351

Paik JH, Ding Z, Narurkar R, Ramkissoon S, Muller F, Kamoun WS, Chae SS, Zheng H, Ying H, Mahoney J, Hiller D, Jiang S, Protopopov A, Wong WH, Chin L, Ligon KL, DePinho RA (2009) FoxOs cooperatively regulate diverse pathways governing neural stem cell homeostasis. Cell Stem Cell 5:540–553

Parker JA, Connolly JB, Wellington C, Hayden M, Dausset J, Neri C (2001) Expanded polyglutamines in Caenorhabditis elegans cause axonal abnormalities and severe dysfunction of PLM mechanosensory neurons without cell death. Proc Natl Acad Sci USA 98:13318–13323

Parker JA, Arango M, Abderrahmane S, Lambert E, Tourette C, Catoire H, Neri C (2005) Resveratrol rescues mutant polyglutamine cytotoxicity in nematode and mammalian neurons. Nat Genet 37:349–350

Pasco MY, Catoire H, Parker JA, Brais B, Rouleau GA, Neri C (2010a) Cross-talk between canonical Wnt signaling and the sirtuin-FoxO longevity pathway to protect against muscular pathology induced by mutant PABPN1 expression in *C. elegans*. Neurobiol Dis 38:425–433

Pasco MY, Rotili D, Altucci L, Farina F, Rouleau GA, Mai A, Neri C (2010b) Characterization of sirtuin inhibitors in nematodes expressing a muscular dystrophy protein reveals muscle cell and behavioral protection by specific sirtinol analogues. J Med Chem 53:1407–1411

Patel KR, Scott E, Brown VA, Gescher AJ, Steward WP, Brown K (2011) Clinical trials of resveratrol. Ann NY Acad Sci 1215:161–169

Renault VM, Rafalski VA, Morgan AA, Salih DA, Brett JO, Webb AE, Villeda SA, Thekkat PU, Guillerey C, Denko NC, Palmer TD, Butte AJ, Brunet A (2009) FoxO3 regulates neural stem cell homeostasis. Cell Stem Cell 5:527–539

Richard T, Pawlus AD, Iglesias ML, Pedrot E, Waffo-Teguo P, Merillon JM, Monti JP (2011) Neuroprotective properties of resveratrol and derivatives. Ann NY Acad Sci 1215:103–108

Rizki G, Iwata TN, Li J, Riedel CG, Picard CL, Jan M, Murphy CT, Lee SS (2011) The Evolutionarily Conserved Longevity Determinants HCF-1 and SIR-2.1/SIRT1 Collaborate to Regulate DAF-16/FOXO. PLoS Genet 7:e1002235

Schuff M, Siegel D, Bardine N, Oswald F, Donow C, Knochel W (2010) FoxO genes are dispensable during gastrulation but required for late embryogenesis in Xenopus laevis. Dev Biol 337:259–273

Schuster E, McElwee JJ, Tullet JM, Doonan R, Matthijssens F, Reece-Hoyes JS, Hope IA, Vanfleteren JR, Thornton JM, Gems D (2010) DamID in *C. elegans* reveals longevity-associated targets of DAF-16/FoxO. Mol Syst Biol 6:399

Siegrist SE, Haque NS, Chen CH, Hay BA, Hariharan IK (2010) Inactivation of both Foxo and reaper promotes long-term adult neurogenesis in Drosophila. Curr Biol 20:643–648

Smith JJ, Kenney RD, Gagne DJ, Frushour BP, Ladd W, Galonek HL, Israelian K, Song J, Razvadauskaite G, Lynch AV, Carney DP, Johnson RJ, Lavu S, Iffland A, Elliott PJ, Lambert PD, Elliston KO, Jirousek MR, Milne JC, Boss O (2009) Small molecule activators of SIRT1 replicate signaling pathways triggered by calorie restriction in vivo. BMC Syst Biol 3:31

Soerensen M, Dato S, Christensen K, McGue M, Stevnsner T, Bohr VA, Christiansen L (2010) Replication of an association of variation in the FOXO3A gene with human longevity using both case–control and longitudinal data. Aging Cell 9:1010–1017

Tank EM, Rodgers KE, Kenyon C (2011) Spontaneous age-related neurite branching in Caenorhabditis elegans. J Neurosci 31:9279–9288

Tissenbaum HA, Guarente L (2001) Increased dosage of a sir-2 gene extends lifespan in Caenorhabditis elegans. Nature 410:227–230

Viswanathan M, Guarente L (2011) Regulation of Caenorhabditis elegans lifespan by sir-2.1 transgenes. Nature 477:E1–E2

Walker FO (2007) Huntington's disease. Lancet 369:218–228

Willcox BJ, Donlon TA, He Q, Chen R, Grove JS, Yano K, Masaki KH, Willcox DC, Rodriguez B, Curb JD (2008) FOXO3A genotype is strongly associated with human longevity. Proc Natl Acad Sci USA 105:13987–13992

Wood JG, Rogina B, Lavu S, Howitz K, Helfand SL, Tatar M, Sinclair D (2004) Sirtuin activators mimic caloric restriction and delay ageing in metazoans. Nature 430:686–689

Yen K, Narasimhan SD, Tissenbaum HA (2011) DAF-16/Forkhead box O transcription factor: many paths to a single Fork(head) in the road. Antioxid Redox Signal 14:623–634

Yu RX, Liu J, True N, Wang W (2008) Identification of direct target genes using joint sequence and expression likelihood with application to DAF-16. PLoS One 3:e1821

Yuan Z, Lehtinen MK, Merlo P, Villen J, Gygi S, Bonni A (2009) Regulation of neuronal cell death by MST1-FOXO1 signaling. J Biol Chem 284:11285–11292

Zuccato C, Valenza M, Cattaneo E (2010) Molecular mechanisms and potential therapeutical targets in Huntington's disease. Physiological reviews 90:905–981

The interplay between the chaperonin TRiC and N-terminal region of Huntingtin mediates Huntington's Disease aggregation and pathogenesis

Koning Shen and Judith Frydman

Abstract Huntington's Disease (HD) is a neurodegenerative disorder resulting from an expanded polyglutamine (polyQ) repeat in exon1 of the Huntingtin (Htt) protein. This polyQ expansion causes aggregation of the Htt protein in neuronal cells, which is linked to HD pathogenesis. Recent evidence has shown that Htt aggregation and toxicity are not solely dictated by the polyQ-expanded region but also by sequences flanking the polyQ region, particularly the first N-terminal 17 amino acids of Htt (N17). N17 has been shown to be critical in the Htt aggregation mechanism as well as host many post-translational modifications and interactions with molecular chaperones. Understanding how N17 functions in Htt aggregation and as a general handle for protein quality control will guide design of HD therapeutics.

Introduction

Huntington's Disease (HD) is an inherited neurodegenerative disorder for which no cure or effective treatment currently exists. Clinical symptoms include chorea, psychiatric abnormalities, and neuronal atrophy primarily in the striatum and cortex (Laccone et al. 1999; Orr and Zoghbi 2007; MacDonald and Gusella 1996). HD brain tissue characteristically contains amyloid aggregates (Davies et al. 1997; Difiglia et al. 1997; Becher et al. 1998), which comprise of an N-terminal fragment of mutant Huntingtin (Htt) protein (Sharp et al. 1995; Difiglia et al. 1995; Duyao et al. 1995; Mangiarini et al. 1996; MacDonald et al. 1993). The Htt mutation linked to HD is the expansion of a CAG-trinucleotide repeat in the first exon of the protein, which manifests as an extended glutamine tract (herein polyQ) near the N-terminus of the Htt protein (MacDonald et al. 1993). In non-HD afflicted individuals, the

K. Shen • J. Frydman (✉)
Department of Biology and BioX Program, Stanford University, Stanford, CA 94305-5020, USA
e-mail: jfrydman@stanford.edu

R.I. Morimoto and Y. Christen (eds.), *Protein Quality Control in Neurodegenerative Diseases*, Research and Perspectives in Alzheimer's Disease,
DOI 10.1007/978-3-642-27928-7_10, © Springer-Verlag Berlin Heidelberg 2013

length of this polyQ varies between 6 to 34 glutamines, while in HD afflicted individuals the length of the polyQ tract is 37 residues and beyond (Orr and Zoghbi 2007). The length of the polyQ tract is tightly correlated with aggregation propensity of the protein *in vitro* and *in vivo* and with earlier age of HD onset and severity of disease phenotype (Orr and Zoghbi 2007), thus linking HD to the so-called aggregation or amyloid diseases.

Understanding Huntingtin aggregation: Looking beyond the polyQ tract

While an extended polyQ tract clearly influences Htt aggregation and toxicity, regions flanking this polyQ tract also exhibit strong effects on aggregation and toxicity. This was first suggested by experiments in yeast which demonstrated that altering flanking sequences of the polyQ region of Htt by introducing artificial tags can strongly influence aggregation and toxicity, for both benign and toxic polyQ lengths (Duennwald et al. 2006). In addition, other protein pathological polyQ disorders besides HD are also linked with extension of a polyQ tract, but the toxic length of this extended tract as well as disease phenotype varies (Orr and Zoghbi 2007). No sequence homology exists among the polyQ proteins of these disorders beyond the polyQ tract (Orr and Zoghbi 2007), suggesting that the differential disease phenotypes may lie in polyQ-flanking sequences. Finally, although polyQ length correlates inversely with age of onset in HD, there is still remarkable diversity in disease pathology even for patients with the same polyQ length (Brinkman et al. 1997; Wexler et al. 2004). These studies suggest that an understanding of HD pathogenesis requires investigation beyond the polyQ tract of the Huntingtin protein.

N17, the N-terminal polyQ flanking sequence of Htt, acts as an "on switch" for aggregation

The full-length Htt protein is a large 350-kDa protein, however, the HD mutation lies within only the polyQ tract of exon1 (MacDonald et al. 1993). Thus, a large proportion of HD studies have focused on characterizing Htt exon1 (Fig. 1A). The two most characterized flanking polyQ sequences in exon1 are the first 17 N-terminal residues (herein N17) and the proline-rich domain (polyPro) immediately C-terminal to the polyQ region. Broadly speaking, the N17 region has been shown to enhance aggregation while the polyPro region has been shown to reduce aggregation (Tam et al. 2009; Thakur et al. 2009; Bhattacharyya et al. 2006; Darnell et al. 2007; Darnell et al. 2009). A large body of work has concentrated on N17 because of its demonstrated role in enhancing Htt aggregation, hosting post-translational modifications and molecular chaperone interactions, and as a potential target for developing anti-HD therapeutics.

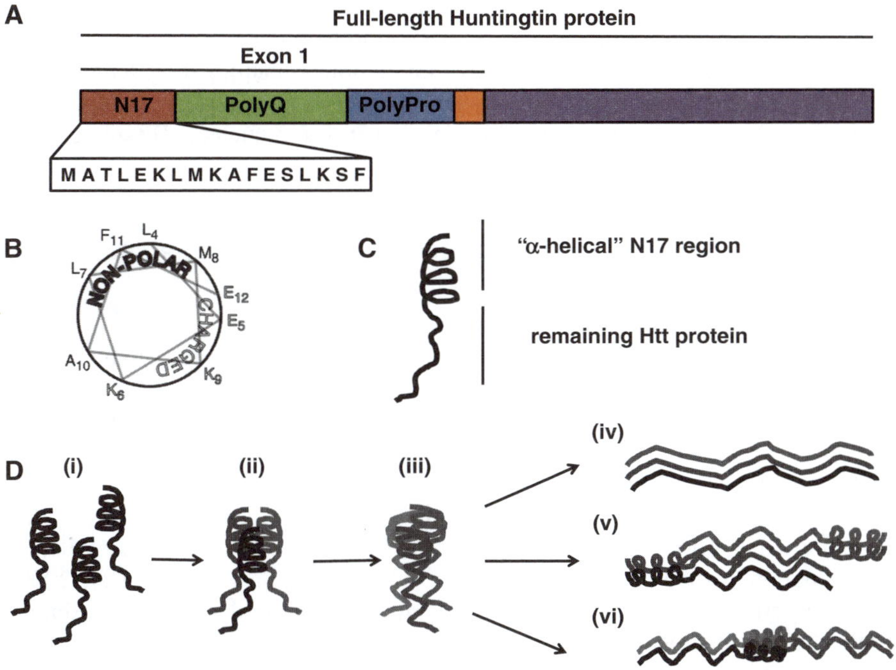

Fig. 1 *The N17 sequence in Huntingtin plays a key role in protein aggregation.* A) Schematic diagram of the polyQ domain and flanking N17 and polyPro sequences compared to the full length Huntingtin (Htt) protein. Diagram is not drawn to scale as the full-length Htt protein is a 3144-amino acid long protein with a 350-kDa molecular weight, while exon 1 is only around 14 kDa (MacDonald et al. 1993). The amino acid sequence of N17 itself is provided. B) Helical wheel diagram of the predicted helical region within N17 sequence (Tam et al. 2009). C) Schematic secondary-structure diagram of the α-helical N17 domain compared to the rest of the more disordered Htt protein. D) Schematic of Htt protein aggregation models involving the α-helical N17 region. Individual Htt monomers (i) oligomerize via the α-helical region to form α-helical oligomers (ii). High local concentration of the polyQ region facilitates β-sheet formation (iii). Three models for the subsequent steps are: (iv) β-sheet formation propagates through the α-helical structure to form β-rich fibrils; (v) propagates β-sheet structure as the α-helical N17 regions become water-exposed; or (vi) α-helical N17 becomes stabilized within the core of the β-rich fibrils

Although aggregation propensity of polyQ proteins is highly correlated with length of the polyQ tract, the presence of the N17 region in Htt variants greatly influences the kinetics of polyQ protein aggregation for both benign and pathogenic polyQ lengths. Deletion of the N17 region in mutant Htt decreases aggregation rates compared to intact mutant Htt (Tam et al. 2009). Addition of the N17 sequence *in trans* rapidly increases the aggregation propensity of pathogenic-length polyQ sequences alone. The mechanism by which N17 increases polyQ aggregation propensity and especially that for mutant Htt is unclear. One possibility is that the N17 region acts to initiate polyQ aggregation. Evidence that the N17 region has binding affinity to both polyQ tracts and the N17 sequence itself suggests that N17 may tether Htt monomers together to nucleate aggregation (Tam et al. 2009).

In addition, N17 may act as a conformational "on switch" to nucleate Htt aggregation, and the "on" conformation may only be promoted in the presence of extended polyQ tracts. Indeed, molecular dynamics simulations suggest that the length of the polyQ tract influences the conformational state of N17 and that conformational changes within N17 serve as key structural transitions to nucleate Htt aggregation (Williamson et al. 2010; Kelley et al. 2009).

N17 enhances Htt aggregation through an α-helical structure

While a solution structure of Htt or Htt-exon1 remains elusive, a number of studies point to an α-helical conformation of the N17 region and suggest this helical property may enhance Htt aggregation. Molecular modeling and circular dichroism studies of the N17 sequence show that N17 forms an amphipathic α-helical structure (Tam et al. 2009; Thakur et al. 2009; Rockabrand et al. 2006; Kelley et al. 2009; Atwal et al. 2007) (Fig. 1B).

While NMR data suggests that the N17 does not contain a highly stable α-helical structure, interactions between different N17 sequences stabilize the α-helical conformation perhaps to further nucleate Htt aggregation (Thakur et al. 2009). A crystal structure of Htt exon1 with 17 glutamines fused to Maltose Binding Protein shows that the N17 region forms an α-helix that also extends into the initial portion of the polyQ region (Kim et al. 2009). Of note, both "helix-breaking" mutations that introduce prolines into the N17 sequence as well as alanine substitutions on the hydrophobic face of the amphipathic helix drastically inhibit mutant Htt aggregation propensity (Tam et al. 2009; Atwal et al. 2007), suggesting that the amphipathic α-helical nature of N17 is critical for Htt aggregation.

A number of structural models have been proposed for how the α-helical nature of N17 participates in the Htt aggregation mechanism (Fig. 1C, D) In one model, the α-helical N17 region forms coiled-coiled domains to mediate Htt oligomerization and subsequently amyloidogenic aggregation. Mutagenesis studies show that modulating the coiled-coil propensity of the polyQ region in Htt can dictate aggregation and toxicity in yeast for both pathogenic and benign polyQ-lengths (Fiumara et al. 2010). While it may seem contradictory to include a helical intermediate in a β-sheet fibrillation pathway, helical oligomers have been proposed as intermediates to β-sheet formation for many other amyloidogenic proteins (Abedini and Raleigh 2009). Here, the α-helical N17 may form coiled-coil oligomers to increase the local concentration of the polyQ tract, which has high β-sheet amyloidogenic propensity. The polyQ β-sheet structure then propagates through the helices to generate the β-sheet amyloid fibril.

In another model, the α-helicity of N17 nucleates the initial oligomerization and β-sheet formation, but N17 retains its α-helical nature by becoming more water-exposed during the aggregation process (Sivanandam et al. 2011; Jayaraman et al. 2012). Solid state NMR data suggests that in the Htt amyloid fibril the N17 region exhibits helical structure and is both mobile and water-exposed (Sivanandam et al. 2011). Here, interactions among N17 regions of different Htt monomers stabilize their α-helical conformations and promote formation of Htt oligomers to nucleate aggregation. Subsequently, during amyloidogenesis, the α-helical N17 regions become more water-exposed and eventually deposit outside of the β-sheet rich amyloid fibril (Sivanandam et al. 2011; Jayaraman et al. 2012).

In a final model, the N17 region retains its α-helicity during amyloidogenesis but becomes tightly packed within the core of the amyloid fibril. EPR spectra of Htt fibrils suggests that the N17 region becomes more structured and involved in tertiary and quaternary interactions packing interactions within the β-sheet fibril to stabilize the fibril core (Bugg et al. 2012). Given the heterogeneity of Htt oligomer and aggregate states, these distinct models could co-exist and lead to distinct types of oligomeric, prefibrillar and fibrillar forms. Nonetheless, these models all share an important role for the α-helical nature of N17 in enhancing aggregation. Outstanding areas of interest include how a polyQ-extended tract influences the aggregation-promoting conformations of N17 and how the structure of an α-helical domain may progress to generate a highly β-sheet rich structure.

N17 modulates Htt localization and is subject to post-translational modifications

The N17 region is highly conserved in evolution and appears to be highly regulated. Several studies have shown that N17 is subject to post-translational modifications *in vivo*, which may be linked to cellular mechanisms regulating Htt quality control. These modifications include SUMOylation, ubiquitination, acetylation, and phosphorylation and exhibit a range of effects on Htt aggregation and toxicity (Thompson et al. 2009; Gu et al. 2009; Steffan 2004; Aiken et al. 2009). Recent work on post-translational modifications has focused on the effect of phosphorylation of two serine residues, Ser13 and Ser16, in the N17 region on Htt toxicity and aggregation (Thompson et al. 2009; Gu et al. 2009; Havel et al. 2011). Phosphorylation of these serine residues in wild-type Htt activates degradation pathways through the proteasome or lysosome (Thompson et al. 2009). However, mutant Htt with an expanded polyQ region is much less efficiently phosphorylated, perhaps suggesting a mechanism of toxicity for mutant Htt (Thompson et al. 2009). Interestingly, phosphomimetic mutations at these serine residues profoundly decrease neurodegenerative pathology in mouse models and protein aggregation *in vitro*, whereas phosphoresistant mutations at these residues do not (Gu et al. 2009). In addition, introducing these phosphomimetic mutations in a mouse striatal cell-line expressing mutant Htt increases cell survival while phosphoresistant mutations decrease cell survival (Atwal et al. 2011). These studies suggest that phosphorylation within N17 decreases HD pathogenesis through either Htt aggregation or cellular protein quality control mechanisms (Thompson et al. 2009; Atwal et al. 2011; Greiner and Yang 2011). How an extended polyQ tract influences efficacy of phosphorylation and other post-translational modifications remains unclear.

In addition, the N17 region has been implicated in the regulation of sub-cellular localization of Htt (Rockabrand et al. 2006; Atwal et al. 2007). For instance, the N17 region has been implicated as an important cytoplasmic retention signal (Steffan 2004) and has been shown to tether Htt to ER and mitochondrial membranes, likely through the amphipathic α-helical region of N17 (Rockabrand et al. 2006; Atwal et al. 2007). Interestingly, post-translational modifications within the N17 region also strongly influence Htt sub-cellular localization. Phosphorylation of the Ser13 and Ser16 residues increases the nuclear localization of both wild-type and mutant Htt (Thompson et al. 2009; Havel et al. 2011; Atwal et al. 2011). This nuclear localization may also play a crucial role in regulating cellular toxicity. Phosphomimetic mutations at Ser13/Ser16 in mutant Htt increase nuclear localization and cell survival, even in the presence of kinase inhibitors (Thompson et al. 2009; Atwal et al. 2011). On the other hand, phosphoresistant mutations at Ser 13/Ser16 decreased mutant Htt nuclear localization and increased cell death (Atwal et al. 2011). Notably, subjecting cells to heat-shock greatly increased nuclear localization of phosphorylated-N17, suggesting that phosphorylation of N17 and nuclear localization of Htt may occur generally in response to cellular stress (Atwal et al. 2011). Since the sub-cellular localization of mutant Htt has been shown to influence toxicity (Saudou et al. 1998; Schilling et al. 1999; Hackam et al. 1999; Cooper et al. 1998), an understanding of how these post-translational modifications influence N17 targeting as well as mutant Htt aggregation and toxicity will be essential in understanding proteostasis regulation of mutant Htt.

The chaperonin TRiC/CCT modulates Htt aggregation and toxicity through association with N17

Molecular chaperones are cellular factors that control protein folding and quality control (Hartl et al. 2011). Chaperones characteristically fold non-native proteins and stabilize non-native polypeptide states against aggregation. Beyond folding, these molecular chaperones also maintain protein quality control by shuttling misfolded proteins to the ubiquitin-proteasome or lysosome system for degradation (Muchowski and Wacker 2005).

Because of the relationship between protein misfolding and protein aggregation and disease (Balch et al. 2008; Chiti and Dobson 2006), recent interest has focused on the role of molecular chaperones in the cellular response to polyQ protein misfolding, aggregation, and toxicity (Muchowski and Wacker 2005). The role of these molecular chaperones in age-related neurological disorders is especially interesting since age-related decline in proteostasis may shift the delicate balance in protein quality control towards pathogenic aggregation (Balch et al. 2008). For instance, the molecular chaperones Hsp70 and Hsp40 have been demonstrated to interact with mutant Htt to suppress aggregation and toxicity (Lotz et al. 2010; Behrends et al. 2006). One molecular chaperone, the ring-shaped chaperonin TRiC/CCT, was recently identified as a regulator of polyQ-aggregation (Behrends

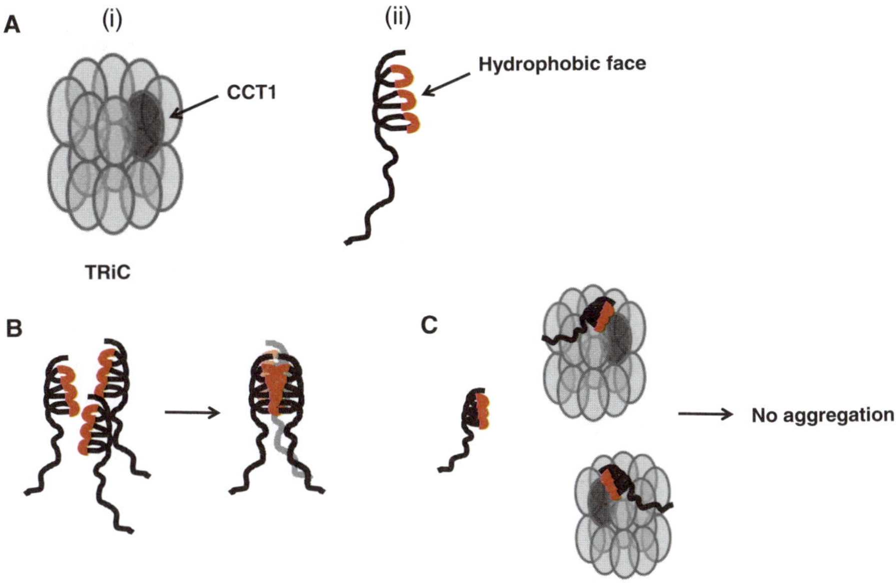

Fig. 2 *The TRiC chaperonin binds to N17 via the hydrophobic α-helical face to block aggregation.* A) (i) Schematic diagram of the TRiC chaperonin with the CCT1 subunit highlighted (Tam et al. 2009). TRiC consists of two stacked rings of eight subunits (CCT1-CCT8) each (Leroux and Hartl 2000). For simplicity, only one CCT1 subunit is highlighted here. (ii) Schematic diagram of the Htt protein with the hydrophobic face of the α-helix highlighted in red. (B) The hydrophobic face may be involved in the interaction of the N17 α-helices in initiating the aggregation mechanism. C) TRiC binds to this hydrophobic face of N17 to block its ability to interact with other N17 regions, thereby preventing accumulation of Htt aggregation

et al. 2006; Kitamura et al. 2006; Tam et al. 2006). TRiC is an essential, hetero-oligomeric chaperonin which consists of eight diverged subunits, CCT1-CCT8, arranged in a double-ring architecture (Leroux and Hartl 2000; Spiess et al. 2004; Yam et al. 2008) (Fig. 2A). The different subunits of TRiC have been shown to have different substrate specificities, one of which (CCT1) can directly bind Htt (Kitamura et al. 2006; Tam et al. 2006; Spiess et al. 2006). Finally, a recent proteomic Htt interactome study using brain tissue of BACHD mice expressing mutant Htt confirmed the *in vivo* interaction between Htt and various subunits of TRiC, including CCT1, indicating that this chaperone is an endogenous modulator of Htt in the brain (Shirasaki et al. 2012). *In vivo* and *in vitro* evidence have shown that TRiC as well as the CCT1 subunit alone can suppress mutant Htt aggregation and toxicity (Behrends et al. 2006; Kitamura et al. 2006; Tam et al. 2006). Knockdown of TRiC or its subunits increases Htt aggregation and toxicity in yeast and mammalian cell culture models, while overexpression of TRiC or CCT1 decreases Htt aggregation and toxicity (Kitamura et al. 2006; Tam et al. 2006). TRiC, in combination with Hsp70, as found to shift the population of Htt aggregates away from smaller, more toxic oligomers to larger, benign oligomeric aggregate species (Behrends et al. 2006). In addition, *in vitro* assays show that TRiC suppresses Htt aggregation most efficiently in the early stages of aggregation (Tam et al. 2006). These data suggest that TRiC may interact with toxic soluble oligomers in the early stages of aggregation to either change the conformation of these oligomers to more benign forms or prevent fibrillation. Synaptic activity has been shown to regulate expression of TRiC, thereby inducing a TRiC-dependent attenuation of mutant Htt aggregation and neurotoxicity (Okamoto et al. 2009).

This finding suggests that TRiC levels could be modulated pharmacologically to protect affected tissues.

The substrate binding sites of each individual CCT subunit reside within the apical domains. Consistent with Htt as a substrate of TRiC/CCT, the interaction between the CCT1 subunit and Htt lies between the substrate-binding domain of CCT1 and the N17 region of Htt (Tam et al. 2009). Furthermore, the binding site on N17 has been localized to the hydrophobic face of the amphipathic helix of N17 (Tam et al. 2009). Since the hydrophobic face of N17 has been shown to be a key enhancer of Htt aggregation propensity (Tam et al. 2009), TRiC may be able to suppress Htt aggregation by binding to this hydrophobic face of N17 and thus blocking amyloidogenesis (Fig. 2C). It is tempting to propose a model where the hydrophobic face of the amphipathic α-helix acts as the conformational switch to nucleate Htt aggregation, (Fig. 2B) and TRiC binding suppresses aggregation by blocking this site. Regardless of the precise mechanism of TRiC-mediated suppression of Htt toxicity, the substrate-specificity of individual TRiC subunits makes taking advantage of the CCT1-N17 interaction an interesting approach for developing HD therapeutics (Spiess et al. 2006).

Conclusion

While the pathogenic mutation in the Htt protein lies within the expansion of the polyQ region, much evidence indicates the importance of looking beyond the polyQ tract and especially at the N17 region to understand the mechanism of Htt aggregation and toxicity. The α-helical nature of N17 provides strong structural clues for how this region could nucleate protein aggregation. In addition, N17 is targeted by multiple post-translational modifications that affect Htt quality control and proteotoxicity. Finally, understanding the specific interaction between N17 and the chaperonin TRiC/CCT harbors great potential for novel approaches to develop a cure for HD.

References

Abedini A, Raleigh DP (2009) A critical assessment of the role of helical intermediates in amyloid formation by natively unfolded proteins and polypeptides. Protein Eng Des Sel 22:453–459

Aiken CT, Steffan JS, Guerrero CM, Khashwji H, Lukacsovich T, Simmons D, Purcell JM, Menhaji K, Zhu YZ, Green K, Laferla F, Huang L, Thompson LM, Marsh JL (2009)

Phosphorylation of threonine 3: implications for Huntingtin aggregation and neurotoxicity. Jo Biol Chem 284:29427–29436

Atwal RS, Xia J, Pinchev D, Taylor J, Epand RM, Truant R (2007) Huntingtin has a membrane association signal that can modulate huntingtin aggregation, nuclear entry and toxicity. Hum Mol Genet 16:2600–2615

Atwal RS, Xia J, Pinchev D, Taylor J, Epand RM, Truant R (2011) Kinase inhibitors modulate huntingtin cell localization and toxicity. Nature Chem Biol 7:453–452

Balch WE, Morimoto RI, Dillin A, Kelly JW (2008) Adapting Proteostasis for Disease Intervention. Science 319:916–919

Becher MW, Kotzuk JA, Sharp AH, Davies SW, Bates GP, Price DL, Ross CA (1998) Intranuclear neuronal inclusions in Huntington's disease and dentatorubral and pallidoluysian atrophy: correlation between the density of inclusions and IT15 CAG triplet repeat length. Neurobiol Dis 4:387–397

Behrends C, Langer CA, Boteva R, Böttcher UM, Stemp MJ, Schaffar G, Rao BV, Giese A, Kretzschmar H, Siegers K, Hartl FU (2006) Chaperonin TRiC promotes the assembly of polyQ expansion proteins into nontoxic oligomers. Mol Cell 23:887–897

Bhattacharyya A, Thakur AK, Chellgren VM, Thiagarajan G, Williams AD, Chellgren BW, Creamer TP, Wetzel R (2006) Oligoproline effects on polyglutamine conformation and aggregation. J Mol Biol 355:524–535

Brinkman RR, Mezei MM, Theilmann J, Almqvist E, Hayden MR (1997) The likelihood of being affected with Huntington disease by a particular age, for a specific CAG size. Am J Hum Genet 60:1202–1210

Bugg CW, Isas JM, Fischer T, Patterson PH, Langen R (2012) Structural Features and Domain Organization of Huntingtin Fibrils. Biol Chem 287:1739–31746

Chiti F, Dobson C (2006) Protein misfolding, functional amyloid, and human disease. Annu Rev Biochem 75:333

Cooper JK, Schilling G, Peters MF, Herring WJ, Sharp AH, Kaminsky Z, Masone J, Khan FA, Delanoy M, Borchelt DR, Dawson VL, Dawson TM, Ross CA (1998) Truncated N-terminal fragments of huntingtin with expanded glutamine repeats form nuclear and cytoplasmic aggregates in cell culture. Hum Mol Genet 7:783–790

Darnell G, Orgel JPRO, Pahl R, Meredith SC (2007) Flanking Polyproline Sequences Inhibit β-Sheet Structure in Polyglutamine Segments by Inducing PPII-like Helix Structure. J Mol Biol 374:688–704

Darnell GD, Derryberry J, Kurutz JW, Meredith SC (2009) Mechanism of Cis-Inhibition of PolyQ Fibrillation by PolyP: PPII Oligomers and the Hydrophobic Effect. Biophys J 97:2295–2305

Davies SW, Turmaine M, Cozens BA, DiFiglia M, Sharp AH, Ross CA, Scherzinger E, Wanker EE, Mangiarini L, Bates GP (1997) Formation of neuronal intranuclear inclusions underlies the neurological dysfunction in mice transgenic for the HD mutation. Cell 90:537–548

Difiglia M, Sapp E, Chase K, Schwarz C, Meloni A, Young C, Martin E, Vonsattel JP, Carraway R, Reeves SA, Boyce FM, Aronin N (1995) Huntingtin is a cytoplasmic protein associated with vesicles in human and rat brain neurons. Neuron 14:1075–1081

Difiglia M, Sapp E, Chase KO, Davies SW, Bates GP, Vonsattel JP, Aronin N (1997) Aggregation of Huntingtin in Neuronal Intranuclear Inclusions and Dystrophic Neurites in Brain. Science 277:1990–1993

Duennwald ML, Jagadish S, Muchowski PJ, Lindquist S (2006) Flanking sequences profoundly alter polyglutamine toxicity in yeast. Proc Natl Acad Sci USA 103:11045–11050

Duyao M, Auerbach AB, Ryan A, Persichetti F, Barnes GT, McNeil SM, Ge P, Vonsattel JP, Gusella JF, Joyner AL, Joyner AL, MacDonald ME (1995) Inactivation of the mouse Huntington's disease gene homolog Hdh. Science 269:407–410

Fiumara F, Fioriti L, Kandel ER, Hendrickson WA (2010) Essential Role of Coiled Coils for Aggregation and Activity of Q/N-Rich Prions and PolyQ Proteins. Cell 143:1121–1135

Greiner ER, Yang XW (2011) Huntington's disease: Flipping a switch on huntingtin. Nat Chem Biol 7:412–414

Gu X, Greiner ER, Mishra R, Kodali R, Osmand A, Finkbeiner S, Steffan JS, Thompson LM, Wetzel R, Yang XW (2009) Serines 13 and 16 Are Critical Determinants of Full-Length Human Mutant Huntingtin Induced Disease Pathogenesis in HD Mice. Neuron 64:828–840

Hackam AS, Singaraja R, Zhang T, Gan L, Hayden MR (1999) In vitro evidence for both the nucleus and cytoplasm as subcellular sites of pathogenesis in Huntington's disease. Hum Mol Genet 8:25–33

Hartl FU, Bracher A, Hayer-Hartl M (2011) Molecular chaperones in protein folding and proteostasis. Nature 475:324–332

Havel LS, Wang CE, Wade B, Huang B, Li S, Li XJ (2011) Preferential accumulation of N-terminal mutant huntingtin in the nuclei of striatal neurons is regulated by phosphorylation. Hum Mol Genet 20:1424–1437

Jayaraman M, Kodali R, Sahoo B, Thakur AK, Mayasundari A, Mishra R, Peterson CB, Wetzel R (2012) Slow Amyloid Nucleation via α-Helix-Rich Oligomeric Intermediates in Short Polyglutamine-Containing Huntingtin Fragments. J Mol Biol 415:881–899

Kelley NW, Huang X, Tam S, Spiess C, Frydman J, Pande VS (2009) The Predicted Structure of the Headpiece of the Huntingtin Protein and Its Implications on Huntingtin Aggregation. J Mol Biol 388:919–927

Kim MW, Chelliah Y, Kim SW, Otwinowski Z, Bezprozvanny I (2009) Secondary Structure of Huntingtin Amino-Terminal Region. Structure 17:1205–1212

Kitamura A, Kubota H, Pack CG, Matsumoto G, Hirayama S, Takahashi Y, Kimura H, Kinjo M, Morimoto RI, Nagata K (2006) Cytosolic chaperonin prevents polyglutamine toxicity with altering the aggregation state. Nat Cell Biol 8:1163–1169

Laccone F, Engel U, Holinski-Feder E, Weigell-Weber M, Marczinek K, Nolte D, Morris-Rosendahl DJ, Zühlke C, Fuchs K, Weirich-Schwaiger H, Schlüter G, von Beust G, Vieira-Saecker AM, Weber BH, Riess O (1999) DNA analysis of Huntington's disease: five years of experience in Germany, Austria, and Switzerland. Neurology 53:801–806

Leroux MR, Hartl FU (2000) Protein folding: versatility of the cytosolic chaperonin TRiC/CCT. Curr Biol 10:R260–4

Lotz GP, Legleiter J, Aron R, Mitchell EJ, Huang SY, Ng C, Glabe C, Thompson LM, Muchowski PJ (2010) Hsp70 and Hsp40 Functionally Interact with Soluble Mutant Huntingtin Oligomers in a Classic ATP-dependent Reaction Cycle. Biol Chem 285:38183–38193

MacDonald ME, Ambrose CM, Duyao MP, Myers RH, Lin C, Srinidhi L, Barnes G, Taylor SA, James M, Groot N, MacFarlane H, Jenkins B, Anderson MA, Wexler NS, Gusella JF, Bates GP, Baxendale S, Hummerich H, Kirby S, North M, Youngman S, Mott R, Zehetner G, Sedlacek Z, Poustka A, Frischauf AM, Lehrach H, Buckler AJ, Church D, Doucette-Stamm L, O'Donovan MC, Riba-Ramirez L, Shah M, Stanton VP, Strobel SA, Draths KM, Wales JL, Dervan P, Housman DE, Altherr M, Shiang R, Thompson L, Fielder T, Wasmuth JJ, Tagle D, Valdes J, Elmer L, Allard M, Castilla L, Swaroop M, Blanchard K, Collins FS, Snell R, Holloway T, Gillespie K, Datson N, Shaw D, Harper PS (1993) A novel gene containing a trinucleotide repeat that is expanded and unstable on Huntington's disease chromosomes. Cell 72:971–983

MacDonald ME, Gusella JF (1996) Huntington's disease: translating a CAG repeat into a pathogenic mechanism. Curr Opin Neurobiol 6:638–643

Mangiarini L, Sathasivam K, Seller M, Cozens B, Harper A, Hetherington C, Lawton M, Trottier Y, Lehrach H, Davies SW, Bates GP (1996) Exon 1 of the HD gene with an expanded CAG repeat is sufficient to cause a progressive neurological phenotype in transgenic mice. Cell 87:493–506

Muchowski PJ, Wacker JL (2005) Modulation of neurodegeneration by molecular chaperones. Nat Rev Neurosci 6:11–22

Okamoto S, Pouladi MA, Talantova M, Yao D, Xia P, Ehrnhoefer DE, Zaidi R, Clemente A, Kaul M, Graham RK, Zhang D, Vincent Chen HS, Tong G, Hayden MR, Lipton SA (2009) Balance between synaptic versus extrasynaptic NMDA receptor activity influences inclusions and neurotoxicity of mutant huntingtin. Nature Med 12:1407–1413

Orr HT, Zoghbi HY (2007) Trinucleotide Repeat Disorders. Annu Rev Neurosci 30:575–621
Rockabrand E, Slepko N, Pantalone A, Nukala VN, Kazantsev A, Marsh JL, Sullivan PG, Steffan JS, Sensi SL, Thompson LM (2006) The first 17 amino acids of Huntingtin modulate its subcellular localization, aggregation and effects on calcium homeostasis. Hum Mol Genet 16:61–77
Saudou F, Finkbeiner S, Devys D, Greenberg ME (1998) Huntingtin acts in the nucleus to induce apoptosis but death does not correlate with the formation of intranuclear inclusions. Cell 95:55–66
Schilling G, Becher MW, Sharp AH, Jinnah HA, Duan K, Kotzuk JA, Slunt HH, Ratovitski T, Cooper JK, Jenkins NA, Copeland NG, Price DL, Ross CA, Borchelt DR (1999) Intranuclear inclusions and neuritic aggregates in transgenic mice expressing a mutant N-terminal fragment of huntingtin. Hum Mol Genet 8:397–407
Sharp AH, Loev SJ, Schilling G, Li SH, Li XJ, Bao J, Wagster MV, Kotzuk JA, Steiner JP, Lo A, Hedreen J, Sisodia S, Snyder SH, Dawson TM, Ryugo DK, Ross CA (1995) Widespread expression of Huntington's disease gene (IT15) protein product. Neuron 14:1065–1074
Shirasaki DI, Greiner ER, Al-Ramahi I, Gray M, Boontheung P, Geschwind DH, Botas J, Coppola G, Horvath S, Loo JA, Yang XW (2012) Network Organization of the Huntingtin Proteomic Interactome in Mammalian Brain. Neuron 75:41–57
Sivanandam VN, Jayaraman M, Hoop CL, Kodali R, Wetzel R, van der Wel PC (2011) The aggregation-enhancing huntingtin N-terminus is helical in amyloid fibrils. J Am Chem Soc 133:4558–4566
Spiess C, Meyer AS, Reissmann S, Frydman J (2004) Mechanism of the eukaryotic chaperonin: protein folding in the chamber of secrets. Trends Cell Biol 14:598–604
Spiess C, Miller EJ, McClellan AJ, Frydman J (2006) Identification of the TRiC/CCT substrate binding sites uncovers the function of subunit diversity in eukaryotic chaperonins. Mol Cell 24:25–37
Steffan JS (2004) SUMO Modification of Huntingtin and Huntington's Disease Pathology. Science 304:100–104
Tam S, Geller R, Spiess C, Frydman J (2006) The chaperonin TRiC controls polyglutamine aggregation and toxicity through subunit-specific interactions. Nat Cell Biol 8:1155–1162
Tam S, Spiess C, Auyeung W, Joachimiak L, Chen B, Poirier MA, Frydman J (2009) The chaperonin TRiC blocks a huntingtin sequence element that promotes the conformational switch to aggregation. Nature Struc Mol Biol 16:1279–1285
Thakur AK, Jayaraman M, Mishra R, Thakur M, Chellgren VM, Byeon IJ, Anjum DH, Kodali R, Creamer TP, Conway JF, Gronenborn AM, Wetzel R (2009) Polyglutamine disruption of the huntingtin exon 1 N terminus triggers a complex aggregation mechanism. Nat Struct Mol Biol 16:380–389
Thompson LM, Aiken CT, Kaltenbach LS, Agrawal N, Illes K, Khoshnan A, Martinez-Vincente M, Arrasate M, O'Rourke JG, Khashwji H, Lukacsovich T, Zhu YZ, Lau AL, Massey A, Hayden MR, Zeitlin SO, Finkbeiner S, Green KN, LaFerla FM, Bates G, Huang L, Patterson PH, Lo DC, Cuervo AM, Marsh JL, Steffan JS (2009) IKK phosphorylates Huntingtin and targets it for degradation by the proteasome and lysosome. J Cell Biol 187:1083–1099
Wexler NS, Lorimer J, Porter J, Gomez F, Moskowitz C, Shackell E, Marder K, Penchaszadeh G, Roberts SA, Gayán J, Brocklebank D, Cherny SS, Cardon LR, Gray J, Dlouhy SR, Wiktorski S, Hodes ME, Conneally PM, Penney JB, Gusella J, Cha JH, Irizarry M, Rosas D, Hersch S, Hollingsworth Z, MacDonald M, Young AB, Andresen JM, Housman DE, De Young MM, Bonilla E, Stillings T, Negrette A, Snodgrass SR, Martinez-Jaurrieta MD, Ramos-Arroyo MA, Bickham J, Ramos JS, Marshall F, Shoulson I, Rey GJ, Feigin A, Arnheim N, Acevedo-Cruz A, Acosta L, Alvir J, Fischbeck K, Thompson LM, Young A, Dure L, O'Brien CJ, Paulsen J, Brickman A, Krch D, Peery S, Hogarth P, Higgins DS Jr, Landwehrmeyer B (2004) Venezuelan kindreds reveal that genetic and environmental factors modulate Huntington's disease age of onset. Proc Natl Acad Sci USA 101:3498–3503

Williamson TE, Vitalis A, Crick SL, Pappu RV (2010) Modulation of Polyglutamine Conformations and Dimer Formation by the N-Terminus of Huntingtin. J Mol Biol 396:1295–1309

Yam AY, Xia Y, Lin HT, Burlingame A, Gerstein M, Frydman J (2008) Defining the TRiC/CCT interactome links chaperonin function to stabilization of newly made proteins with complex topologies. Nat Struct Mol Biol 15:1255–1262

Index

R.I. Morimoto and Y. Christen (eds.), *Protein Quality Control in Neurodegenerative Diseases*, Research and Perspectives in Alzheimer's Disease,
DOI 10.1007/978-3-642-27928-7, © Springer-Verlag Berlin Heidelberg 2013

Printing: Ten Brink, Meppel, The Netherlands
Binding: Stürtz, Würzburg, Germany